SRA Connecting Math Concepts

LEVEL C

Columbus, Ohio

www.sra4kids.com

 **SRA
McGraw-Hill**

Send all inquiries to:
SRA/McGraw-Hill
4400 Easton Commons
Columbus, OH 43219

Printed in the United States of America.

ISBN 0-02-684691-8

 9 0 RRC 09 08

The **McGraw-Hill** Companies

Lesson 12

- To find the number of squares in a rectangle, first figure out the numbers for the column arrow and the row arrow.
- Then write the multiplication problem and the answer. Remember to start with the **column** number.

Use lined paper. Write a, b, c and d.

Next to each letter, write the multiplication problem and the answer.

a.

b.

c.

d.

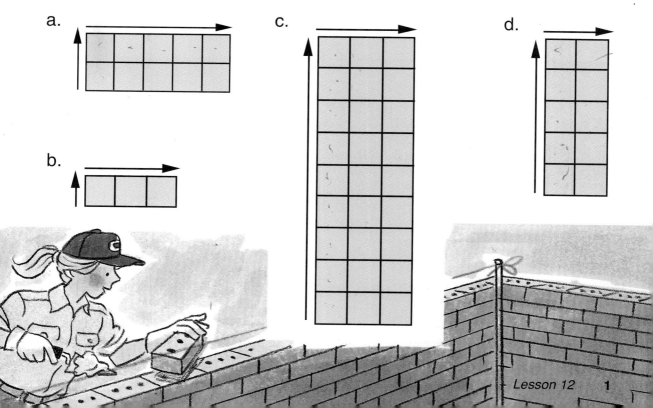

- The missing number in each family is shown with a box.
- If the big number is missing, you add.
- If a small number is missing, you subtract.

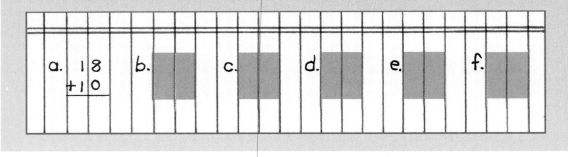

Use lined paper. Write a column problem and the answer for each number family.

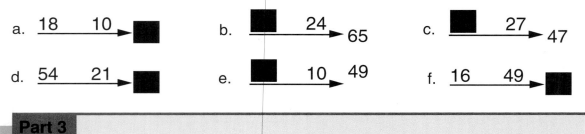

a. 18   10 ▸ ■

b. ■   24 ▸ 65

c. ■   27 ▸ 47

d. 54   21 ▸ ■

e. ■   10   49 ▸

f. 16   49 ▸ ■

Part 3

- You've learned to write two addition facts for number families.
- Both facts start with a small number and end with the big number.
- The first fact goes forward along the arrow.

4   6 ▸ 10

4 + 6 = 10

6 + 4 = 10

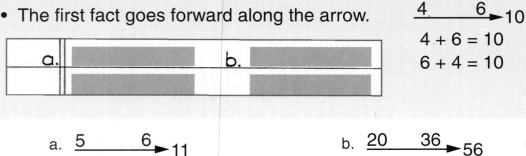

a. 5   6 ▸ 11

b. 20   36 ▸ 56

Lesson 13

- You can't see the squares inside the rectangles, but the numbers on the arrows show how many squares are in each column and how many columns there are.

- Write the multiplication problem and the answer for each rectangle. Remember to start with the column number.

Use lined paper. Write this:

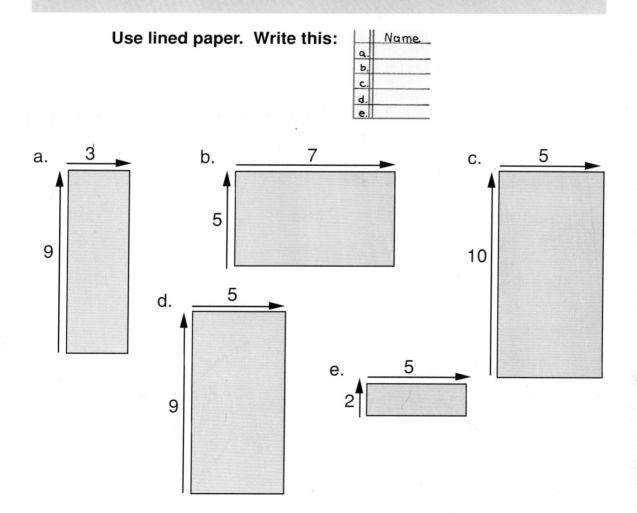

Part 2

- The hour hand is the short hand.

- The minute hand is the long hand.

- To read the hour, start at the top of the clock and go to the number just before the hour hand.

- To figure out the minutes, start at the top of the clock and count by 5 to the minute hand.

- You write the hour first, then a colon, then the minutes.

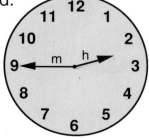

hour   minutes

4:35

a.

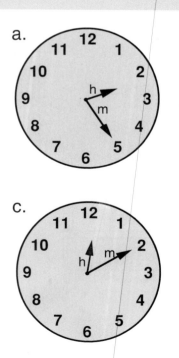

b.

c.

d.

Lesson 14

Part 1 Read the last two digits of each numeral.

a. 576   b. 319   c. 514   d. 791   e. 112

Part 2 Use lined paper. Write this:

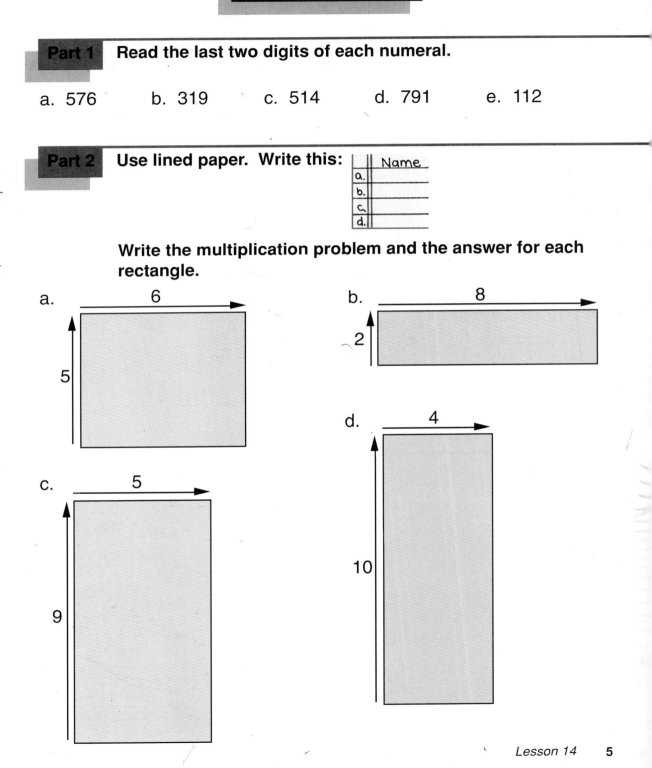

```
   Name
a.
b.
c.
d.
```

Write the multiplication problem and the answer for each rectangle.

a.   6
5

b.   8
2

c.   5
9

d.   4
10

- Read the problem in the ones column. If the problem starts with the smaller number, cross out the tens digit and rewrite the top number.

- If the problem in the ones column starts with the big number, just work the problem. You don't have to borrow.

- Remember, if you borrow you must rewrite the top number.

Use lined paper. Copy each problem and work it.

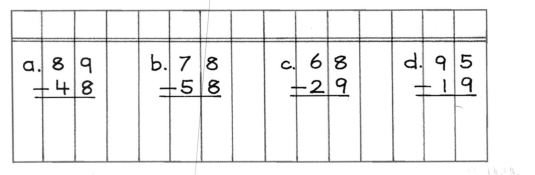

Part 4

a.

b.

c.

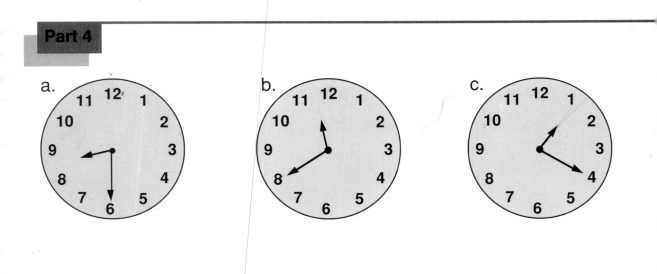

Lesson 15

Part 1 Use lined paper. Write the letters a, b, c, d, e. Write the multiplication problem and the answer for each rectangle.

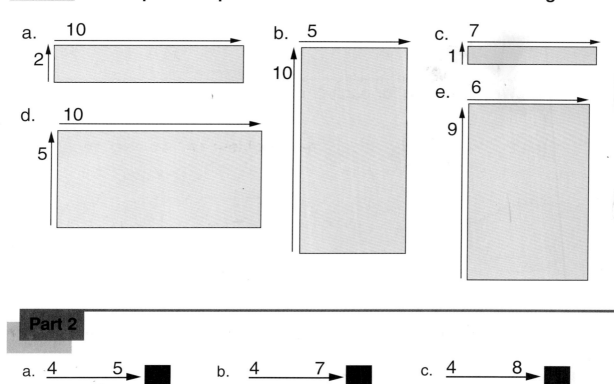

a. 10 2

b. 5 10

c. 7 1

d. 10 5

e. 6 9

Part 2

a. 4   5

b. 4   7

c. 4   8

Use lined paper. Write both addition facts for each number family. Start the first fact with the first small number.

a.	b.	c.

Use lined paper. Write the column problem and the answer
for each number family.

a.
		7	2
	+ 2		1
		9	3

a. 7 ▮ → 39

b. ▮ 12 → 57

c. 2 53 → ▮

d. 20 ▮ → 24

e. 4 19 → ▮

f. 7 ▮ → 10

Part 4 Write the time for each clock.

a.

b.

c.

Part 5 Read the last two digits of each numeral.

a. 436   b. 587   c. 118   d. 513   e. 266

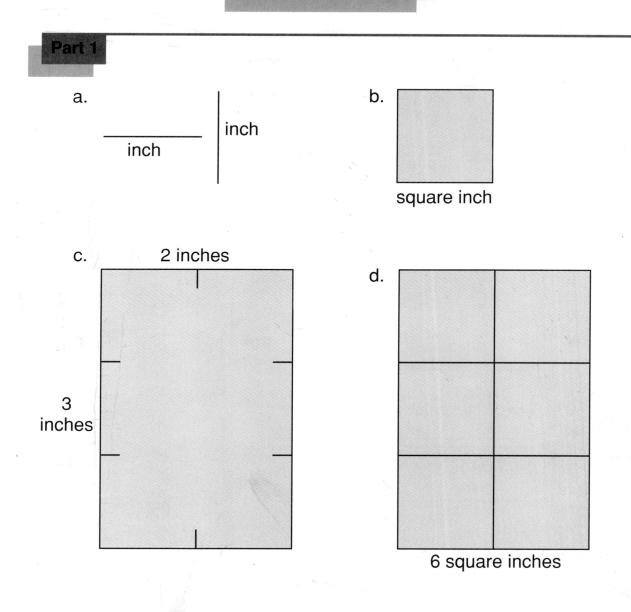

a.

_____ | inch
inch

b.

square inch

c. 2 inches

3 inches

d.

6 square inches

- The answer to area problems is always in **square units.**
- If the units shown are **inches,** the answer to the area problem is in **square inches.**
- If the units are **feet,** the answer is in **square feet.**

Write the multiplication problem and the whole answer.

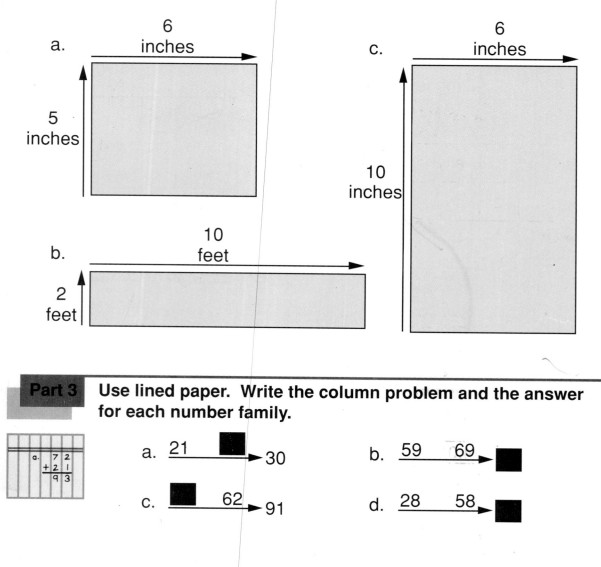

a.

6
inches

5
inches

c.

6
inches

10
inches

b.

10
feet

2
feet

Use lined paper. Write the column problem and the answer for each number family.

a.
	7	2
+	2	1
	9	3

a. 21 ■ ➤ 30

b. 59 69 ➤ ■

c. ■ 62 ➤ 91

d. 28 58 ➤ ■

Lesson 17

Part 1

- The first column shows the hours people worked on Monday.
- The second column shows the hours people worked on Tuesday.
- The third column shows the hours people worked on Wednesday.

	Hours on Monday	Hours on Tuesday	Hours on Wednesday	Total for all days
Dick	4	6	9	19
Mary	2	9	1	12
Harriet	1	1	4	6
Total for all people	7	16	14	

a. How many hours did Dick work on Wednesday?

b. Who worked the most hours on Tuesday?

c. How many hours did all the people work on Monday?

d. How many hours did Harriet work on all three days?

Part 2

Use lined paper.
Write each problem in a column.
Copy the amounts that are shown on the price tags.
Add and write the answer.

```
a.  7 2
  + 2 1
    9 3
```

a. $\boxed{\$ .75}$ $\boxed{\$4.01}$ $\boxed{\$2.00}$

```
$  .75
   4.01
+ 2.00
$ 6.76
```

b. $\boxed{\$2.00}$ $\boxed{\$1.22}$ $\boxed{\$ .04}$

c. $\boxed{\$7.05}$ $\boxed{\$ .31}$ $\boxed{\$1.29}$

d. $\boxed{\$3.54}$ $\boxed{\$ .29}$ $\boxed{\$1.11}$

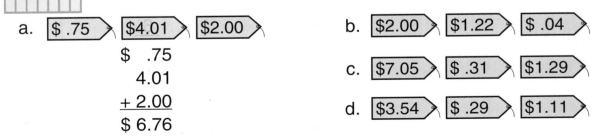

- If a problem says that a person **buys more, finds more** or **gets more,** you add.

  You end up with the big number.

- If a problem says that a person has some and then **sells, loses** or **gets rid of some,** you subtract.

  You end up with a small number.

a.

$$\begin{array}{r} 7\ 2 \\ +\ 2\ 1 \\ \hline 9\ 3 \end{array}$$

a. A scientist saw 17 stars. Then she saw 12 more stars. How many stars did she see in all?

b. A store had 87 suits. The store sells 59 suits. How many suits does the store end up with?

c. Jan had 12 books. She finds 39 more books. How many books does she end up with?

d. Alex had 74 pencils. He sold 45 pencils. How many pencils did he end up with?

Copy each problem and work it.

a. 356   b. 377   c. 347   d. 970
  − 249   − 127   − 228   − 761

a.	7	2
+	2	1
	9	3

Independent Work

Part 5   **Write the time for each clock.**

a.

b.

c.

Do parts 4 and 5 in your workbook.

Part 1 Write the answer to each question.

	Elm Street	Oak Street	Maple Street	Total for all streets
Red cars	4	5	9	18
Yellow cars	2	2	8	12
Blue cars	4	4	1	9
Total for all cars	10	11	18	

a. What's the total number of yellow cars on all the streets?

b. What's the total number of cars that went down Maple Street?

c. Which had more blue cars on it, Elm Street or Maple Street?

d. Which street had the fewest cars?

e. Were there more red cars or blue cars on Oak Street?

Part 2 Write the multiplication problem and the whole answer.

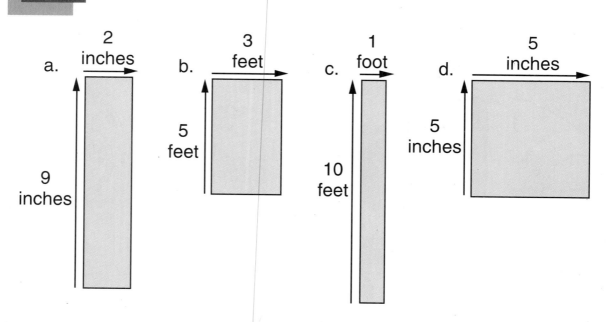

a. 2 inches / 9 inches

b. 3 feet / 5 feet

c. 1 foot / 10 feet

d. 5 inches / 5 inches

Paired Practice

a. 4 + 4 = ■   b. 4 + 5 = ■   c. 4 + 8 = ■   d. 4 + 9 = ■
e. 4 + 10 = ■   f. 4 + 2 = ■   g. 4 + 7 = ■   h. 4 + 1 = ■
i. 4 + 6 = ■   j. 4 + 9 = ■   k. 4 + 4 = ■   l. 4 + 3 = ■

Independent Work

Part 4   **Write the time for each clock.**

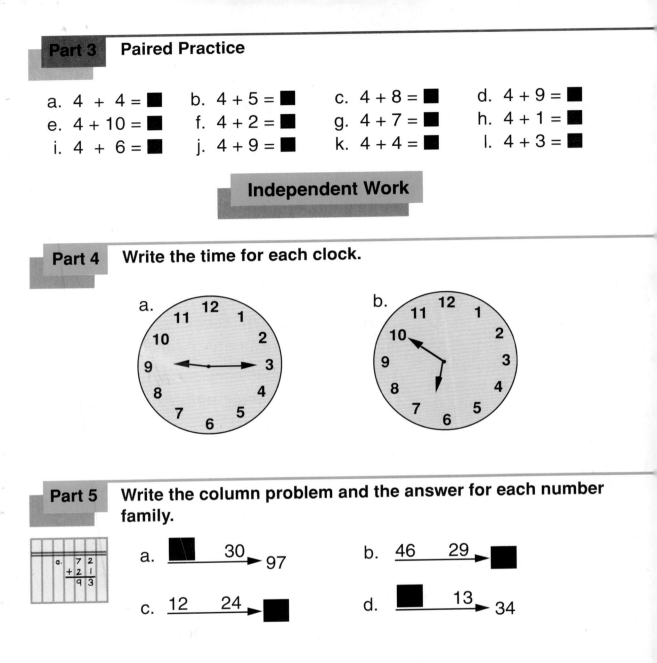

Part 5   **Write the column problem and the answer for each number family.**

a. ■ ──30──▶ 97

b. 46 ──29──▶ ■

c. 12 ──24──▶ ■

d. ■ ──13──▶ 34

Part 1 Write the answer to each question.

This table shows how much rain fell in different cities during May, June and July. The rain is measured in inches.

	May	June	July	Total for all months
River City	6	9	1	16
Hill Town	3	1	8	12
Oak Grove	0	7	9	16
Total for all cities	9	17	18	

a. Which month had the most rainfall?

b. Which month had the least rainfall?

c. Which city had the least amount of rain?

d. How much rain fell in all the cities during July?

Part 2 Write both addition facts for each number family on your lined paper.

a. $\xrightarrow{\quad 5 \qquad 7 \quad}$ ■

b. $\xrightarrow{\quad 5 \qquad 8 \quad}$ ■

Part 3 Write each problem in a column. Remember the decimal point and the dollar sign.

a. $2.36 > $1.11 > $2.20 > b. $3.29 > $1.79 > $2.51 >

c. $3.56 > $ .64 > $5.15 > d. $1.34 > $2.85 > $2.46 >

Part 4 Copy each problem and work it.

a. 355
   − 149

b. 894
   − 685

c. 636
   − 219

Part 5 Write the time for each clock.

Lesson 20

Part 1 **Find the area of each rectangle.**

A mile is over 5,000 feet.

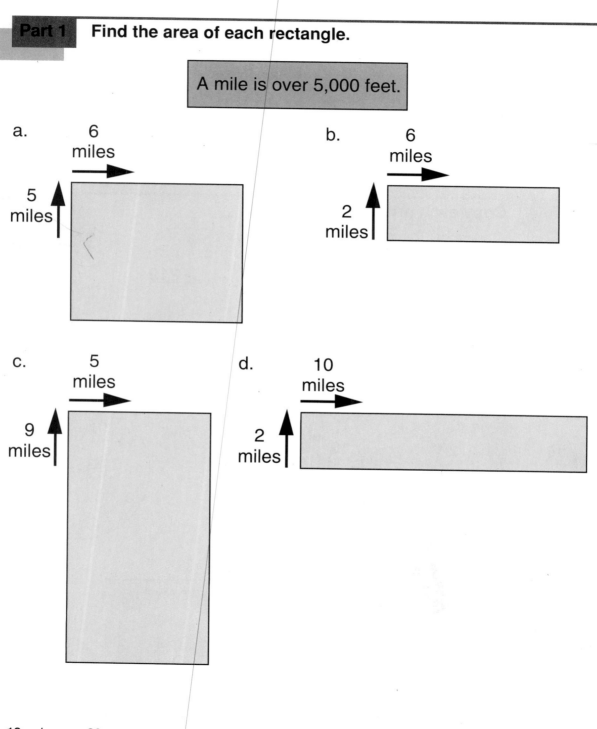

a.
6
miles

5
miles

b.
6
miles

2
miles

c.
5
miles

9
miles

d.
10
miles

2
miles

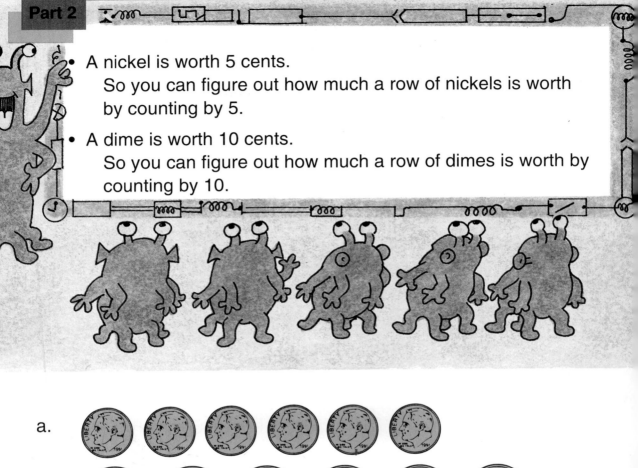

- A nickel is worth 5 cents.
  So you can figure out how much a row of nickels is worth by counting by 5.

- A dime is worth 10 cents.
  So you can figure out how much a row of dimes is worth by counting by 10.

a.

b.

c.

d.

e.

Part 3 For each word problem, write the column-addition or column-subtraction problem and the answer.

a. A dog had 99 fleas. Then the dog got rid of 70 fleas. How many fleas did the dog end up with?

b. A cat had 14 fleas. The cat picked up 44 more fleas. How many fleas did the cat end up with?

c. A tree was 51 feet tall. Then the tree grew 48 more feet. How tall did the tree end up being?

d. A farmer had 56 cows. She sold 29 cows. How many cows did she end up with?

Lesson 21

Part 1

The sign should be big next to the bigger number.
The sign should be small next to the smaller number.

a. 7 ⃔ 6   b. 5 ∠ 9   c. 56 ∠ 59   d. 305 ⃔ 35   e. 20 ∠ 200

Part 2

**Write both addition facts for each number family.
Start the first fact with the first small number.**

a. 5 ———7——▶ ■   b. 5 ———8——▶ ■

Part 3

Copy each problem and write the answer.

a. 5 + 9 = ■   b. 5 + 8 = ■   c. 5 + 7 = ■   d. 5 + 5 = ■

e. 2 + 5 = ■   f. 4 + 5 = ■   g. 5 + 8 = ■   h. 5 + 10 = ■

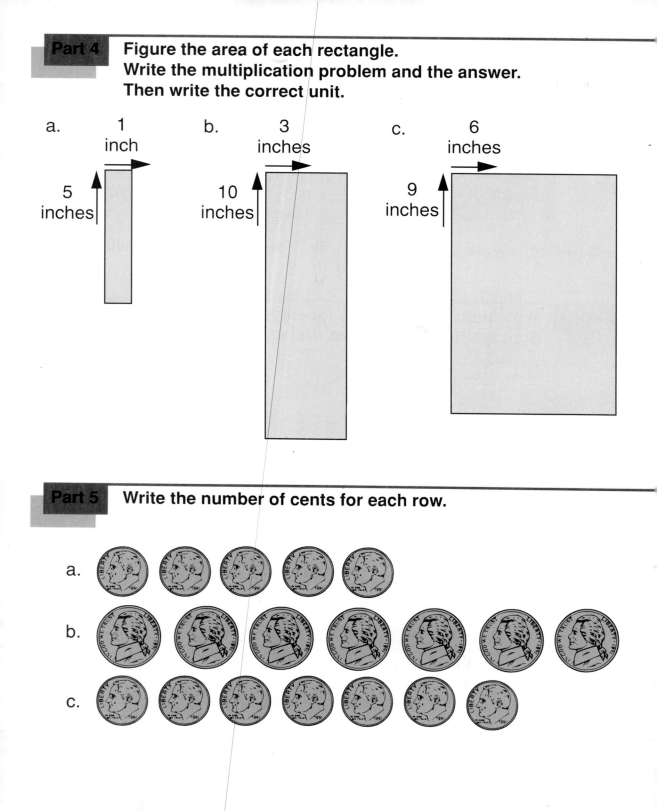

Part 4 Figure the area of each rectangle.
Write the multiplication problem and the answer.
Then write the correct unit.

a.
1
inch

5
inches

b.
3
inches

10
inches

c.
6
inches

9
inches

Part 5 Write the number of cents for each row.

a.

b.

c.

Lesson 22

Part 1 Write the answer to each problem.

a. $9 + 5 = $ ■   b. $8 + 5 = $ ■   c. $7 + 5 = $ ■   d. $6 + 5 = $ ■

e. $5 + 7 = $ ■   f. $4 + 5 = $ ■   g. $5 + 8 = $ ■   h. $10 + 5 = $ ■

Part 2

> Sometimes a letter is used instead of a box in a number family. The letter works just like a box. It's the missing number.

a. The first small number is 14. The second small number is J. The big number is 66.

b. The big number is 89. The first small number is R. The second small number is 28.

c. The second small number is 36. The first small number is 19. The big number is J.

d. The big number is 99. The first small number is P. The second small number is 28.

Part 3 Write both addition facts for each number family. Start the first fact with the first small number.

a. $\xrightarrow{\quad 5 \qquad 8 \quad}$ ■   b. $\xrightarrow{\quad 5 \qquad 7 \quad}$ ■

Write the subtraction fact for each number family.

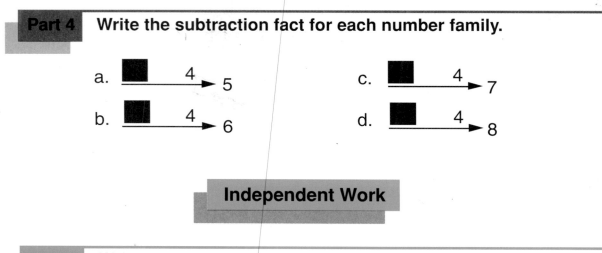

a. ■ ——4——▶ 5

b. ■ ——4——▶ 6

c. ■ ——4——▶ 7

d. ■ ——4——▶ 8

Independent Work

Part 5 **Write the addition problem or the subtraction problem and the answer.**

a. Bill had 236 pounds of sand. He sold 217 pounds of sand. How many pounds of sand did he end up with?

b. Jill had 200 pounds of sand. Then she got 380 more pounds of sand. How many pounds of sand did she end up with?

Part 6 **Write the number of cents for each row.**

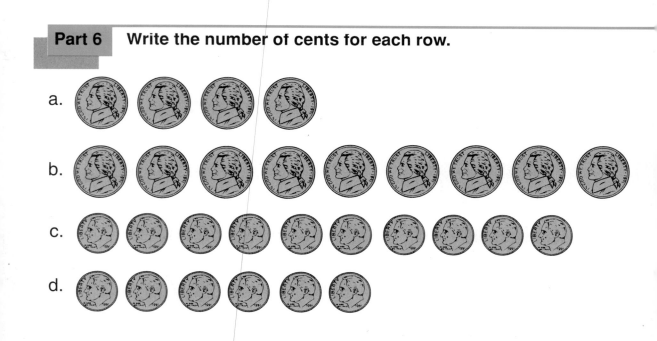

a.

b.

c.

d.

Lesson 23

Part 1

Make the number family. Work the addition problem or the subtraction problem.

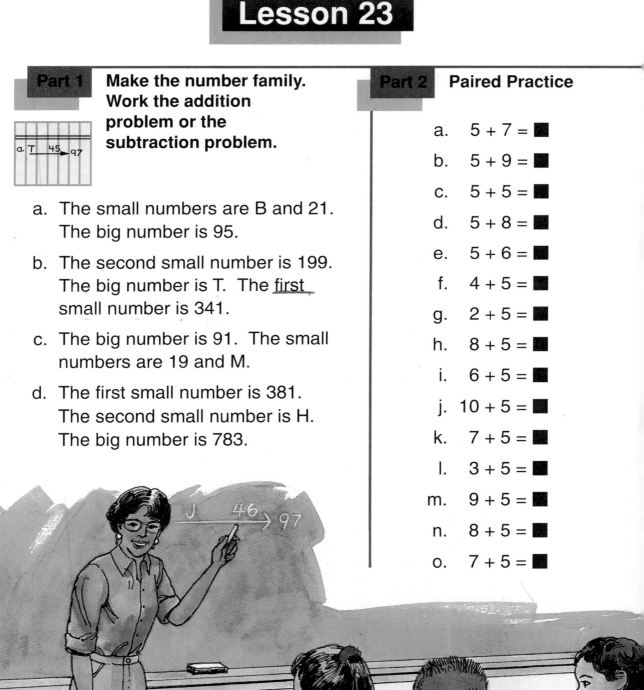

a. | T | 45 ➤ 97

a. The small numbers are B and 21. The big number is 95.

b. The second small number is 199. The big number is T. The <u>first</u> small number is 341.

c. The big number is 91. The small numbers are 19 and M.

d. The first small number is 381. The second small number is H. The big number is 783.

Part 2   Paired Practice

a.   5 + 7 = ■

b.   5 + 9 = ■

c.   5 + 5 = ■

d.   5 + 8 = ■

e.   5 + 6 = ■

f.   4 + 5 = ■

g.   2 + 5 = ■

h.   8 + 5 = ■

i.   6 + 5 = ■

j. 10 + 5 = ■

k.   7 + 5 = ■

l.   3 + 5 = ■

m.   9 + 5 = ■

n.   8 + 5 = ■

o.   7 + 5 = ■

J   46 ➤ 97

Part 3 Copy each problem and work it.

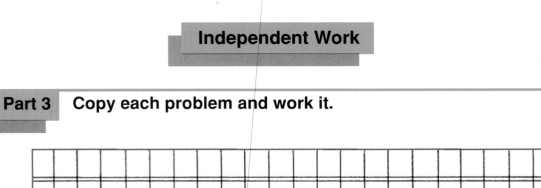

a.
```
  7 6 5
+   7 7
```

b.
```
  3 7 8
+ 4 8 5
```

c.
```
  2 5 7
+ 8 6 5
```

d.
```
  2 9 4
+ 5 4 7
```

Part 4 Write the time shown by each clock.

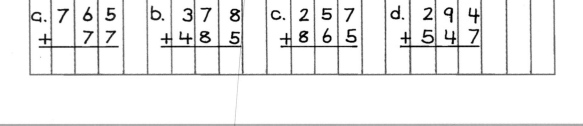

a.

b.

c.

Lesson 24

- A lot of word problems tell about amounts that somebody **spends.** For some of these problems you subtract. For some of these problems you add.

- Here's a subtraction problem:

    A person has 7 dollars and **spends** 5 dollars. How much does the person end up with?

- The problem tells how much money the person has at first and tells that the person spends 5 dollars.

- Here's an addition problem:

    A person **spends** 7 dollars. Then the person **spends** 9 more dollars. How much does the person spend in all?

- The person spends more than 7 dollars. To find how much the person spends, you have to add 7 and 9. The answer is 16 dollars.

a. A person spends 2 dollars and then spends 5 more dollars. How much does the person spend in all?

b. A person has 30 dollars and then spends 20 dollars. How much money does the person end up with?

Part 2

- A person is not buying all the items that are shown with price tags.
- Read each problem to see what the person buys.
- Add those amounts to figure out how much the person spends.

Write the addition problem and the answer.

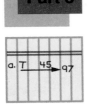

1	2	3	4
$3.45	$2.35	$ .71	$1.69

a. A person buys items 1 and 3. How much does the person spend?

b. A person buys items 2, 3 and 4. How much does the person spend?

c. A person buys items 1, 2 and 4. How much does the person spend?

Part 3

Make the number family.
Work the addition or subtraction problem.
Then cross out the letter in the number family and write the answer.

a. The big number is M. The small numbers are 103 and 749.

b. The second small number is 9. The big number is 84. The first small number is R.

c. The first small number is 264. The second small number is T. The big number is 765.

d. The second small number is 92. The first small number is 138. The big number is B.

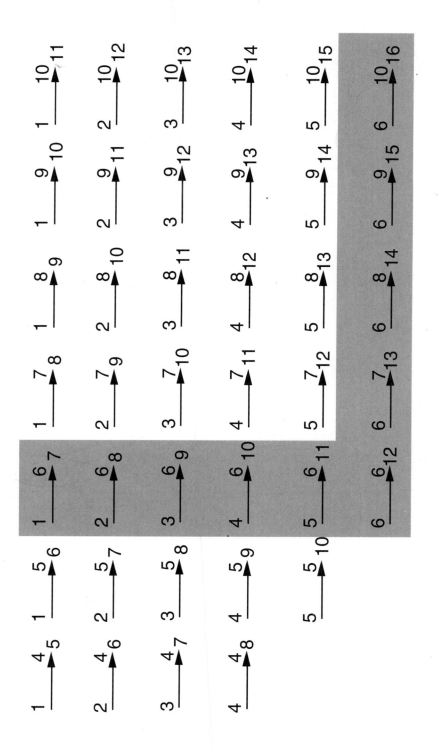

Part 4 Copy each problem and work it.

	a.	258	b.	267	c.	274	d.	105
		21		+ 575		+ 487		445
		+ 664						+ 358

Part 5 Write the time for each clock.

a.

b.

Lesson 25

Part 1
Write both addition facts for each number family. Start the first fact with the first small number.

a.  6   7 ■

b. 6   8 ■

Part 2
For each problem make a number family.
Write the addition or subtraction problem and figure out the answer. Then fix up the number family.

a. The big number is 173. The second small number is 61. The first small number is N.

b. The small numbers are 482 and 18. The big number is M.

c. The first small number is R. The big number is 95. The second small number is 49.

d. The big number is 549. The small numbers are 132 and T.

Part 3
Copy each problem and work it.

a.	b.	c.	d.
352	506	884	876
− 149	− 382	− 290	− 647

Part 4
Copy each problem. Then complete the sign.

a. 5 — 9   b. 30 — 50   c. 301 — 300   d. 500 — 50

- Here are 2 pictures of the same rectangle.
- In the first picture the rectangle is lying flat.
- In the second picture, it is standing up. It's the same rectangle.
- So no matter which side you start with when you figure the area, you'll end up with the same number of squares.

Write the multiplication problem and the answer for each rectangle. Remember the unit name.

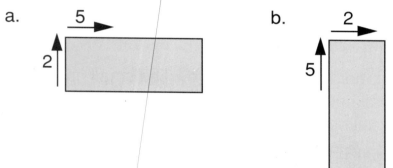

a. 5   2

b. 2   5

Write the multiplication problem and the answer for each rectangle.

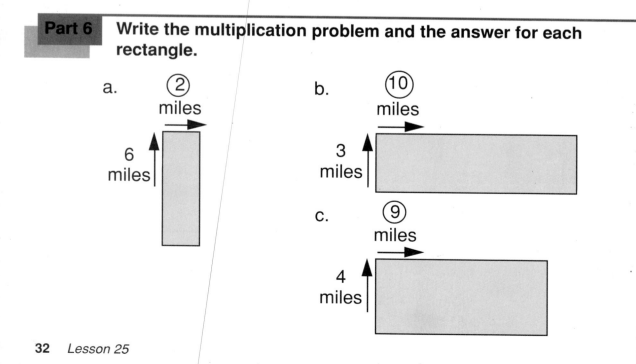

a. ②
    miles
    6
    miles

b. ⑩
    miles
    3
    miles

c. ⑨
    miles
    4
    miles

Part 7 — Copy each problem and work it.

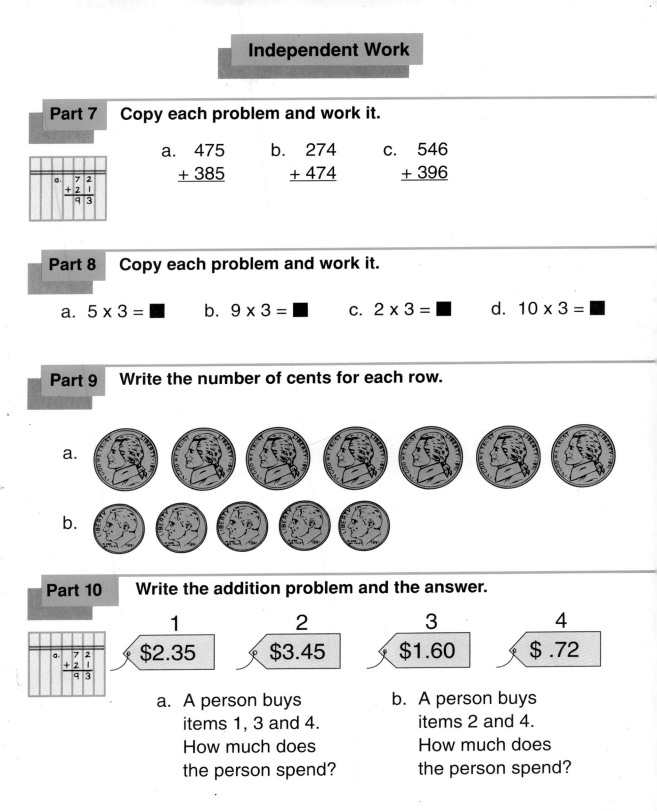

a. 475
   + 385

b. 274
   + 474

c. 546
   + 396

Part 8 — Copy each problem and work it.

a. 5 x 3 = ■   b. 9 x 3 = ■   c. 2 x 3 = ■   d. 10 x 3 = ■

Part 9 — Write the number of cents for each row.

a.

b.

Part 10 — Write the addition problem and the answer.

1	2	3	4
$2.35	$3.45	$1.60	$ .72

a. A person buys items 1, 3 and 4. How much does the person spend?

b. A person buys items 2 and 4. How much does the person spend?

Part 1

- Sometimes we refer to a number without telling which number it is.

- We can call that number **J** or **B** or any other letter.

- Here is a sentence that tells about two numbers:

  T is less than H.

- We don't know which numbers T and H are, but we can put those numbers in a number family.

- T is less than H. So T is a small number.

- H is the big number.

Use lined paper. For each sentence, make a number family and put the two letters in it.

a. J is less than M.　　b. W is bigger than R.　　c. P is larger than T.

d. H is less than T.　　e. Y is more than T.

Part 2 Write the multiplication problem and the answer for each rectangle.
Start with the circled number.
The unit for the answer is square miles.

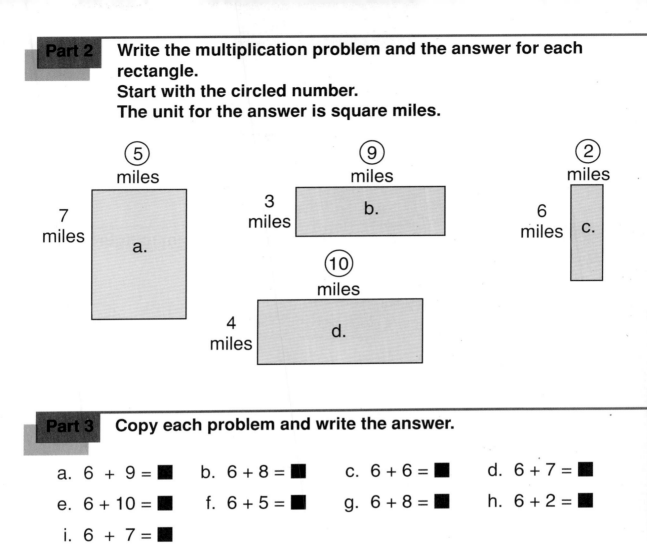

Part 3 Copy each problem and write the answer.

a. $6 + 9 = \blacksquare$ b. $6 + 8 = \blacksquare$ c. $6 + 6 = \blacksquare$ d. $6 + 7 = \blacksquare$

e. $6 + 10 = \blacksquare$ f. $6 + 5 = \blacksquare$ g. $6 + 8 = \blacksquare$ h. $6 + 2 = \blacksquare$

i. $6 + 7 = \blacksquare$

Part 4 For each problem make a number family.
Write the addition or subtraction problem and figure out the answer. Then fix up the number family.

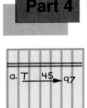

a. The big number is 671. The second small number is R. The first small number is 621.

b. The small numbers are T and 281. The big number is 395.

c. The first small number is 372. The big number is N. The second small number is 529.

Part 5 Copy each problem and work it.

a. 667
  − 458

b. 558
  − 167

c. 638
  − 294

Part 6 For each problem, write the addition problem to figure out how much money each person spends.

1
$2.75

2
$3.70

3
$4.20

4
$2.05

a. A man buys items 1, 2 and 4. How much money does he spend?

b. A girl buys items 2, 3 and 4. How much money does she spend?

c. A man buys items 1, 2 and 3. How much money does he spend?

Do the independent work for Lesson 26, Part 3, in your workbook.

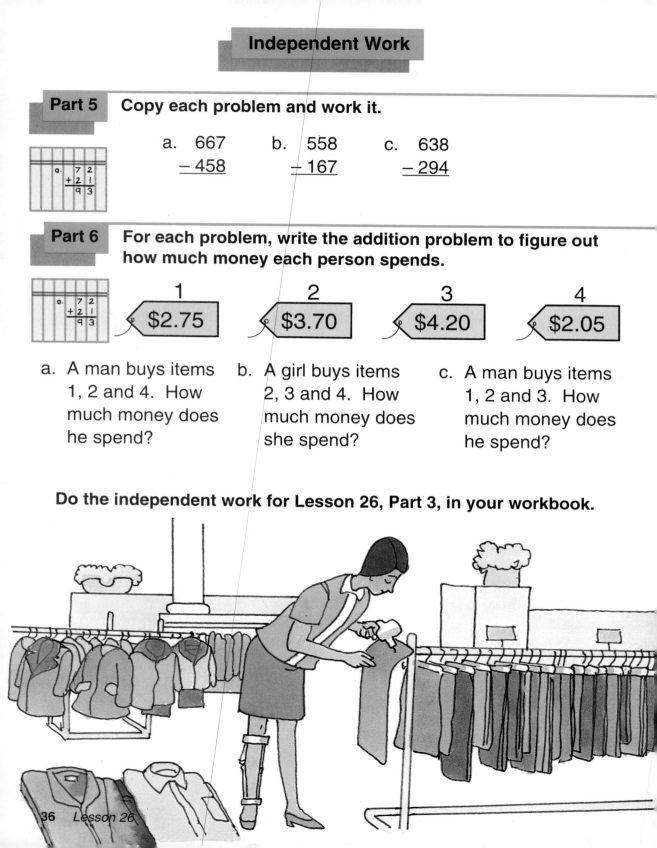

Lesson 27

Part 1   **For each sentence, make a number family and put the two letters in it.**

    a. P is larger than W.   b. F is smaller than T.

    c. Y is more than V.   d. T is bigger than B.

Part 2   **For each problem make a number family.**
Write the addition or subtraction problem and figure out
the answer. Then fix up the number family.

    a. The first small number is 174. The big number is H. The second small number is 694.

    b. The big number is 542. The small numbers are W and 451.

Part 3

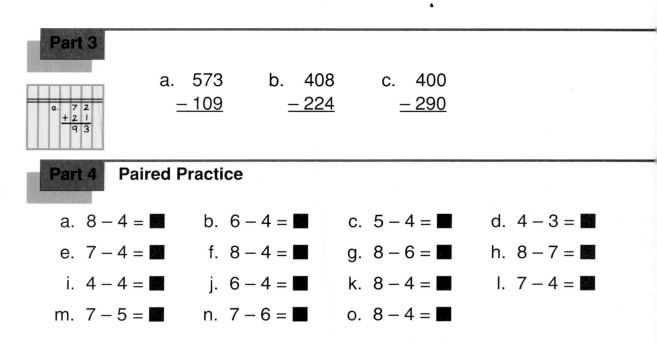

      a. 573   b. 408   c. 400

        $-\,109$ $-\,224$ $-\,290$

Part 4   **Paired Practice**

   a. $8 - 4 = \blacksquare$ b. $6 - 4 = \blacksquare$ c. $5 - 4 = \blacksquare$ d. $4 - 3 = \blacksquare$

   e. $7 - 4 = \blacksquare$ f. $8 - 4 = \blacksquare$ g. $8 - 6 = \blacksquare$ h. $8 - 7 = \blacksquare$

   i. $4 - 4 = \blacksquare$ j. $6 - 4 = \blacksquare$ k. $8 - 4 = \blacksquare$ l. $7 - 4 = \blacksquare$

   m. $7 - 5 = \blacksquare$ n. $7 - 6 = \blacksquare$ o. $8 - 4 = \blacksquare$

Part 5 Write the multiplication problem and the answer for each rectangle.
Remember to start with the circled number.
The unit for each answer is square feet.

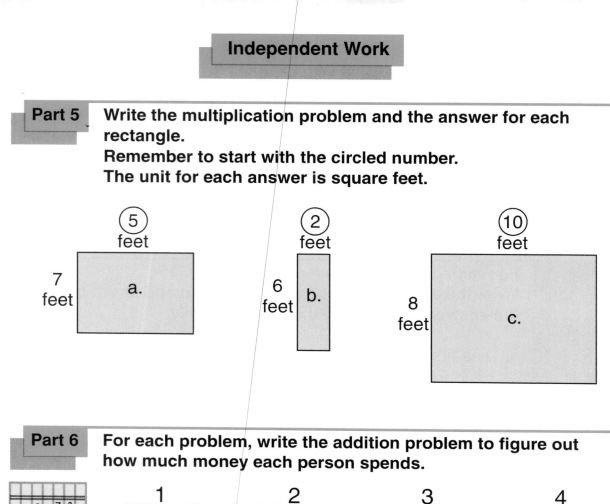

Part 6 For each problem, write the addition problem to figure out how much money each person spends.

1	2	3	4
$6.02	$1.13	$2.80	$3.99

a. A person buys items 1, 3 and 4. How much does the person spend?

b. A person buys items 2, 3 and 4. How much does the person spend?

c. A person buys items 1, 2 and 3. How much does the person spend?

Write the column problems and figure out the answers.

a. A dog had 431 fleas. The dog got rid of 112 fleas. How many fleas did the dog end up with?

b. A dog had 467 fleas. Then the dog got 479 more fleas. How many fleas did the dog end up with?

Don't forget to do the independent work in your workbook.

Lesson 27, Part 3

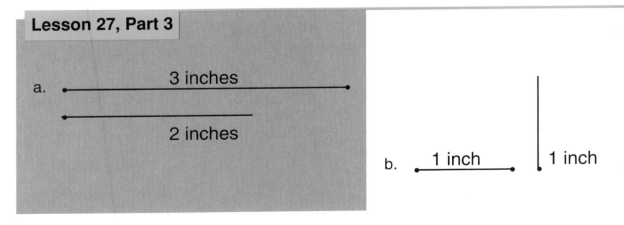

a. 3 inches

2 inches

b. 1 inch   1 inch

Lesson 28

Part 1 Write the subtraction fact for each number family.

a. ⬛ ——5—→ 6

b. ⬛ ——5—→ 7

c. ⬛ ——5—→ 8

d. ⬛ ——5—→ 9

e. ⬛ ——5—→ 10

Part 2

- One of the numbers for each rectangle is a number you know how to count by.
- For some of the rectangles, that number is on the side.
- For some of the rectangles, that number is on the bottom.

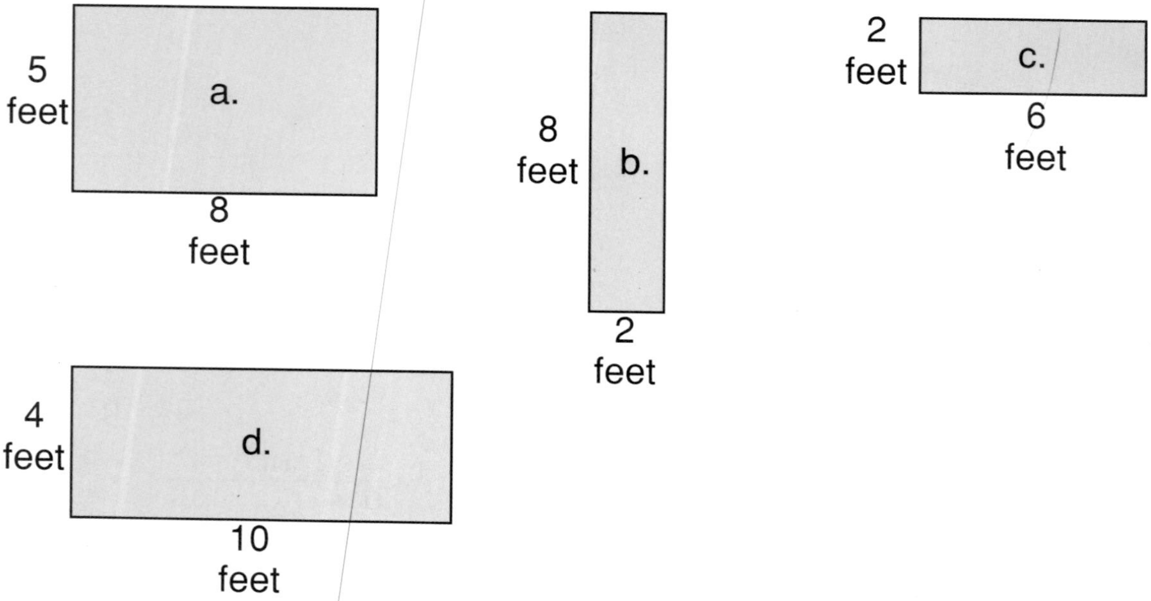

5 feet | a. |
8 feet

8 feet | b. | 2 feet

2 feet | c. | 6 feet

4 feet | d. |
10 feet

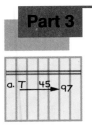

Part 3

For each problem make a number family.
Write the addition or subtraction problem and figure out
the answer. Then fix up the number family.

a. The second small number is R. The big number is 741.
   The first small number is 19.

b. The small numbers are 673 and 38. The big number is M.

Independent Work

Part 4

Copy each problem and work it.

a.	336	b.	336	c.	258	d.	113
	+ 468		63		+ 596		448
			+ 577				+ 364

Lesson 29

Part 1   Make a number family for each problem.
First put the two values in the number family.
Then put the circled number in the family.

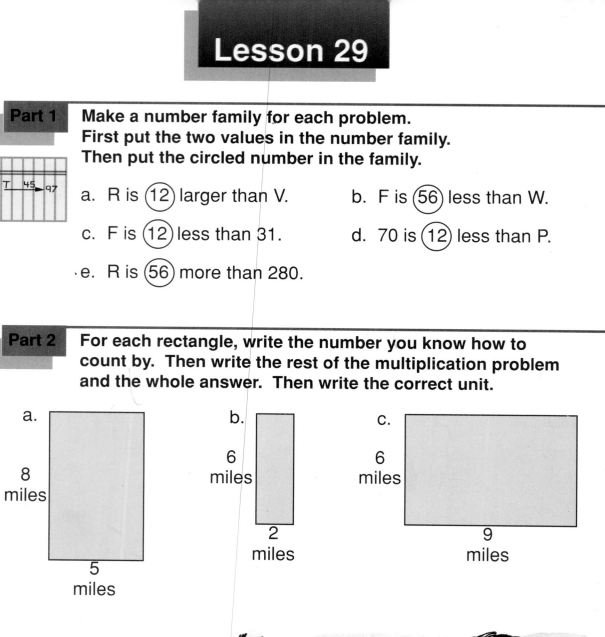

a. T ⟶ 45 ⟶ 97

a. R is ⟨12⟩ larger than V.

b. F is ⟨56⟩ less than W.

c. F is ⟨12⟩ less than 31.

d. 70 is ⟨12⟩ less than P.

·e. R is ⟨56⟩ more than 280.

Part 2   For each rectangle, write the number you know how to count by. Then write the rest of the multiplication problem and the whole answer. Then write the correct unit.

a.

8
miles

5
miles

b.

6
miles

2
miles

c.

6
miles

9
miles

D ∠12 | 14   12   10 |

- Here's a statement about D:

D is less than 12.

So D could be any number that is less than 12.

- D couldn't be 14 because 14 is not less than 12.
- D couldn't be 12 because 12 is not less than 12.
- D could be 10 because 10 is less than 12.

Write the numbers that could be the box or the letter.

a. R ≻17 | 17   10   18   4   56 | b. ☐∠ 21 | 22   21   20   4 |

c. ☐≻3 | 20   5   4   1   0 | d. T∠5 | 4   3   2   1   0 |

Part 4 **Write the number of cents for each row.**

a.

b.

c.

d.

e.

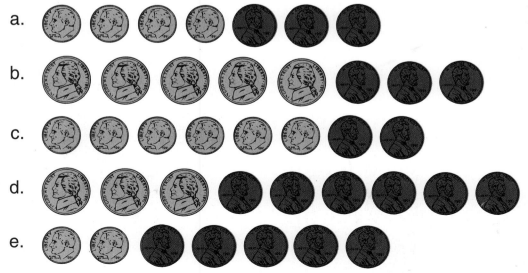

Lesson 30

Part 1

Make a number family for each problem.
Then write the addition or subtraction problem and figure
out what number F is.

a. F is ⑫ more than 56.

b. 96 is ⑰ less than F.

c. 96 is ⑰ more than F.

d. F is ㉗ less than 59.

Part 2

Start with the circled number and write the multiplication
problem and the answer.
Then write the other multiplication problem and the
answer.

When you work area problems, you get the same
answer no matter which side you start with.

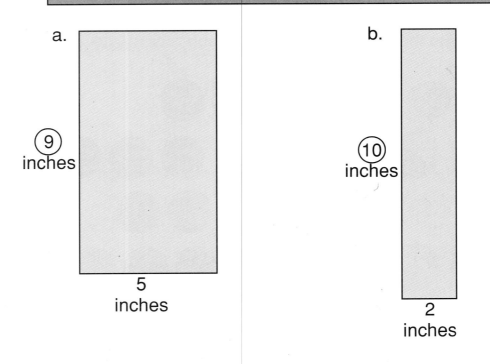

a.

⑨
inches

5
inches

b.

⑩
inches

2
inches

Part 3 Write the number of cents for each row.

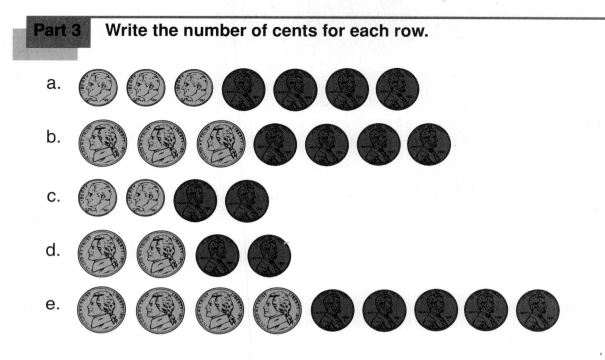

a.

b.

c.

d.

e.

Independent Work

Part 4 Copy the problems and write the answers.

a.		
a.	7	2
+ 2		1
9	3

a.　760　　b.　244　　c.　360　　d.　　70
　　　147　　　　　703　　　　　　49　　　　　983
　　+　　8　　　　+ 422　　　　+ 253　　　　+　　6

Part 5 Write the answer for each problem. Do not copy the problems.

a. 9 x 4 = ■　　b. 9 x 9 = ■　　c. 9 x 3 = ■　　d. 9 x 10 = ■

e. 9 x 5 = ■　　f. 9 x 1 = ■　　g. 5 x 1 = ■　　h. 5 x 9 = ■

i. 1 x 5 = ■　　j. 5 x 6 = ■　　k. 5 x 10 = ■　　l. 5 x 7 = ■

Test 3

Part 5

One of the numbers for each rectangle is a number you know how to count by. Find the area of each rectangle. Remember the unit in the answer.

a.

5 miles

6 miles

b.

4 miles

10 miles

Part 6

Make a number family for each problem.

a. R is 13 more than M.

b. U is 15 less than W.

Part 7

For each problem, make a number family.
Write the addition or subtraction problem and figure out the answer. Then fix up the number family.

a. T ──45──→ 97

a. The small numbers are 482 and 18. The big number is M.

b. The first small number is R. The big number is 95. The second small number is 49.

Part 8

A person is not buying all the items that are shown with the price tags. Read each problem to see what the person buys. Add up only those amounts.

a.
```
  7 2
+ 2 1
  9 3
```

1 — $3.75

2 — $2.35

3 — $ .71

4 — $1.09

a. A person buys items 1 and 3. How much does the person spend?

b. A person buys items 2, 3 and 4. How much does the person spend?

Part 9 Write 1 and 2 on your paper.
Read each description.
Then write the letter for the correct rectangle.

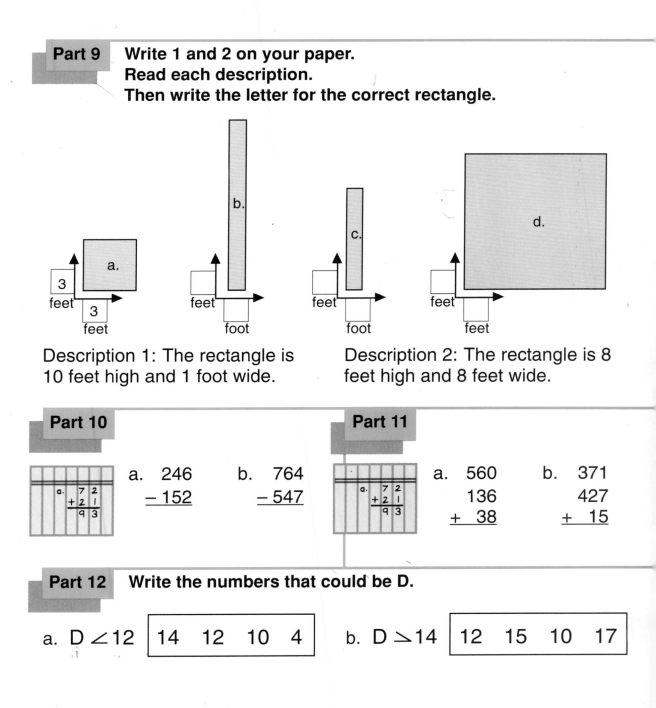

Description 1: The rectangle is 10 feet high and 1 foot wide.

Description 2: The rectangle is 8 feet high and 8 feet wide.

Part 10

a. 246   b. 764
  − 152   − 547

Part 11

a. 560   b. 371
   136   427
+ 38   + 15

Part 12 Write the numbers that could be D.

a. D ∠ 12 | 14   12   10   4 |

b. D ≥ 14 | 12   15   10   17 |

Lesson 31

Part 1   Write two problems and answers for each rectangle.
First write the multiplication problem and the answer that
starts with the circled number.

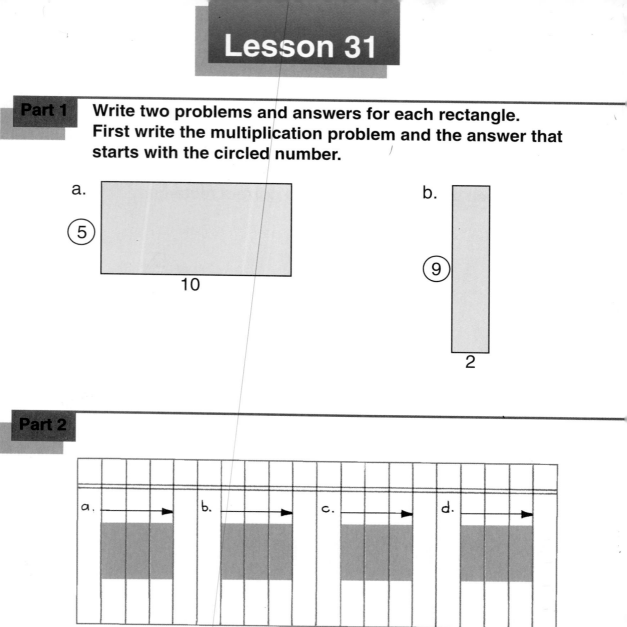

a.

⑤

10

b.

⑨

2

Part 2

a.   b.   c.   d.

a. M is ㊹ more than 17.

c. 239 is ㊵ more than M.

b. 12 is ㉛ less than M.

d. M is ⑲⓪ less than 500.

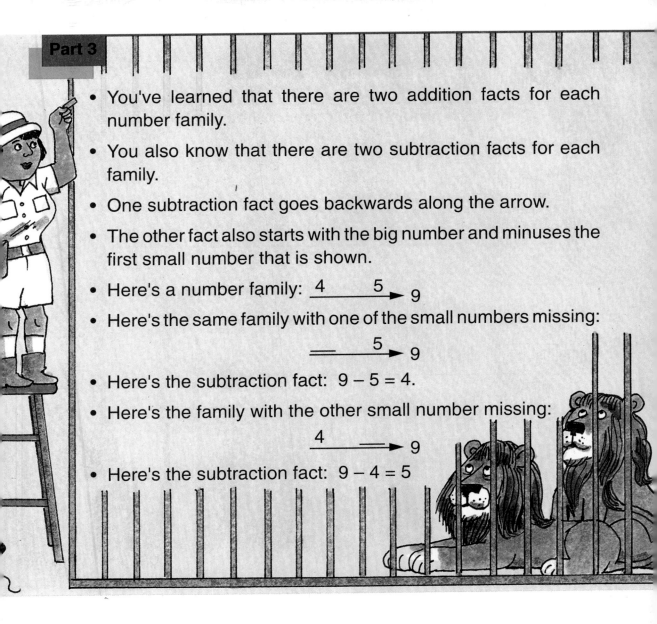

Part 3

- You've learned that there are two addition facts for each number family.
- You also know that there are two subtraction facts for each family.
- One subtraction fact goes backwards along the arrow.
- The other fact also starts with the big number and minuses the first small number that is shown.
- Here's a number family: $\underset{\longrightarrow}{4 \qquad 5}\; 9$
- Here's the same family with one of the small numbers missing:

  $\underset{\longrightarrow}{\underline{\quad\quad} \quad 5}\; 9$

- Here's the subtraction fact: $9 - 5 = 4.$
- Here's the family with the other small number missing:

  $\underset{\longrightarrow}{4 \quad \underline{\quad\quad}}\; 9$

- Here's the subtraction fact: $9 - 4 = 5$

For each family write the subtraction fact that ends with the missing number.

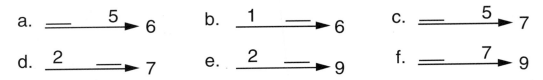

a. $\underset{\longrightarrow}{\underline{\quad\quad} \quad 5}\; 6$

b. $\underset{\longrightarrow}{1 \quad \underline{\quad\quad}}\; 6$

c. $\underset{\longrightarrow}{\underline{\quad\quad} \quad 5}\; 7$

d. $\underset{\longrightarrow}{2 \quad \underline{\quad\quad}}\; 7$

e. $\underset{\longrightarrow}{2 \quad \underline{\quad\quad}}\; 9$

f. $\underset{\longrightarrow}{\underline{\quad\quad} \quad 7}\; 9$

Part 4 Copy each problem and complete the sign.

a. 643 — 634   b. 59 — 509   c. 20 — 199

d. 74 — 17   e. 210 — 201   f. 18 — 21

Lesson 24, Part 2

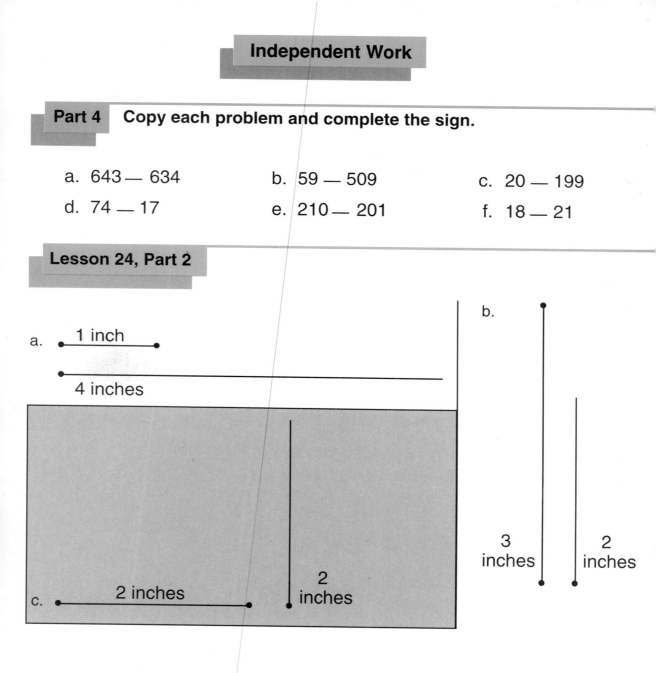

a. 1 inch

4 inches

b.

3 inches   2 inches

c. 2 inches   2 inches

Lesson 32

Part 1

a. $\xlongequal{\quad 9 \quad}$ 11   c. $\xlongequal{\quad 5 \quad}$ 9   e. $\xlongequal{\quad 3 \quad}$ 5

b. $2 \xrightarrow{\qquad}$ 11   d. $4 \xrightarrow{\qquad}$ 9   f. $2 \xrightarrow{\qquad}$ 5

Part 2

- You can make number families for word problems.
- Here's a problem: **56 is ⑭ less than P. What number is P?**
- Here's how you work the problem: You read the first sentence without 14: **56 is less than P.**

  Here's the number family: $\xrightarrow{\quad 56 \quad}$ P

- Now you read the whole sentence: 56 is ⑭ **less than P.**

  You write 14 in the only place it can go: $\underline{\quad 14 \qquad 56 \quad}\rightarrow$ P

- The question asks: **What number is P?**

  To find P you write the addition problem and the answer:

$$\begin{array}{r} 1 \\ 1\,4 \\ +\,5\,6 \\ \hline 7\,0 \end{array}$$

- To show what number P is, cross out P in the number family and write 70: $\underline{\quad 14 \qquad 56 \quad}\rightarrow \overset{70}{\cancel{P}}$

Part 3 For each problem, make the number family.
Then write the addition problem or the subtraction problem
and figure out what number P is.

a. P is ⑳ less than 52. What number is P?

b. P is ㉖ less than 96. What number is P?

c. 59 is ㉛ more than P. What number is P?

d. 42 is ⑥ less than P. What number is P?

e. P is ⑳ more than 490. What number is P?

Part 4 First draw rectangles for each description.
Then write the numbers to show how high and how wide
each rectangle is.
Next to each rectangle write the area problem and the
answer.

a. The rectangle is 2 feet high and 7 feet wide.

b. The rectangle is 5 feet high and 7 feet wide.

Part 5 Write two problems and answers for each rectangle.
First write the multiplication problem and the answer that
starts with the circled number.

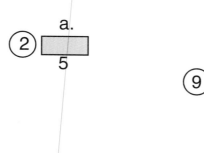

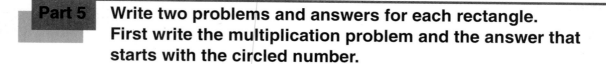

Part 6   Copy each problem and write the answer.

a.	b.	c.	d.
456 − 65	949 − 840	978 − 797	200 − 110

Lesson 33

Part 1   **First draw rectangles for each description.**
Then write the numbers to show how high and how wide
each rectangle is.
Next to each rectangle write the area problem and the
answer.

a. The rectangle is 5 feet high and 7 feet wide.

b. The rectangle is 9 feet high and 10 feet wide.

Part 2

- Number families for multiplication look different from number families for addition.

- The multiplication number families are like area problems.

  Here's an area problem: The rectangle is 2 units high and 5 units wide.

  It has 10 squares inside: 2 [grid 10]

  You can see the arrows for the two sides of the rectangle.

  The number for the up arrow is 2.

  The number for the other arrow is 5.

  There are 10 squares in the rectangle. That's the big number.

- Here's the number family with the same three numbers:

$$2 \overline{}^{\,5}_{} \quad 10$$

  The family looks just like the area problem.

- The number 2 is first. The number 5 is along the top arrow. The big number is 10.

Write two multiplication facts for each number family.
Start the first fact with the first small number.

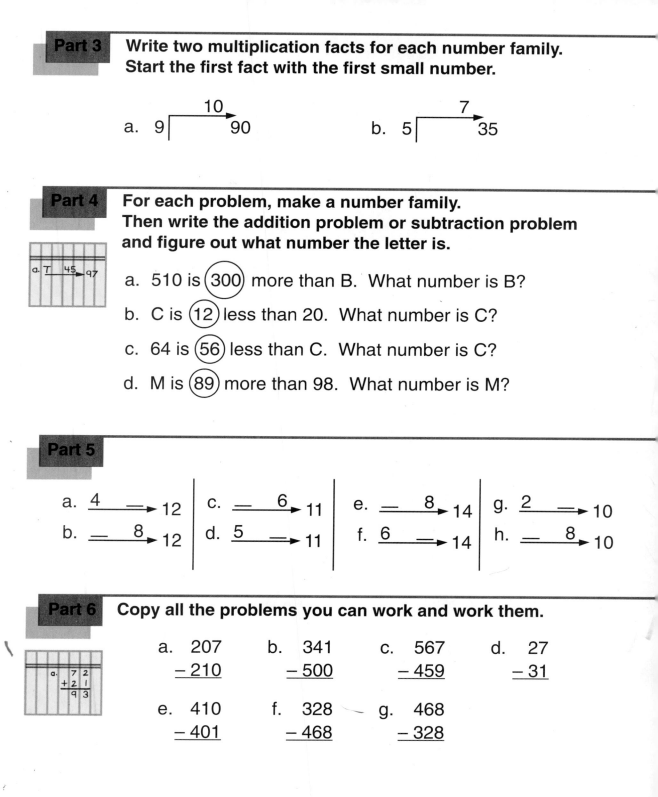

a. 9 ⌐10⟶ 90

b. 5 ⌐7⟶ 35

Part 4 **For each problem, make a number family.**
Then write the addition problem or subtraction problem
and figure out what number the letter is.

a. 510 is (300) more than B. What number is B?

b. C is (12) less than 20. What number is C?

c. 64 is (56) less than C. What number is C?

d. M is (89) more than 98. What number is M?

Part 5

a. 4 ⟶ 12

b. ── 8 ⟶ 12

c. ── 6 ⟶ 11

d. 5 ⟶ 11

e. ── 8 ⟶ 14

f. 6 ⟶ 14

g. 2 ⟶ 10

h. ── 8 ⟶ 10

Part 6 **Copy all the problems you can work and work them.**

a. 207
  − 210

b. 341
  − 500

c. 567
  − 459

d. 27
  − 31

e. 410
  − 401

f. 328
  − 468

g. 468
  − 328

Independent Work

Part 7 Write the number of cents for each row.

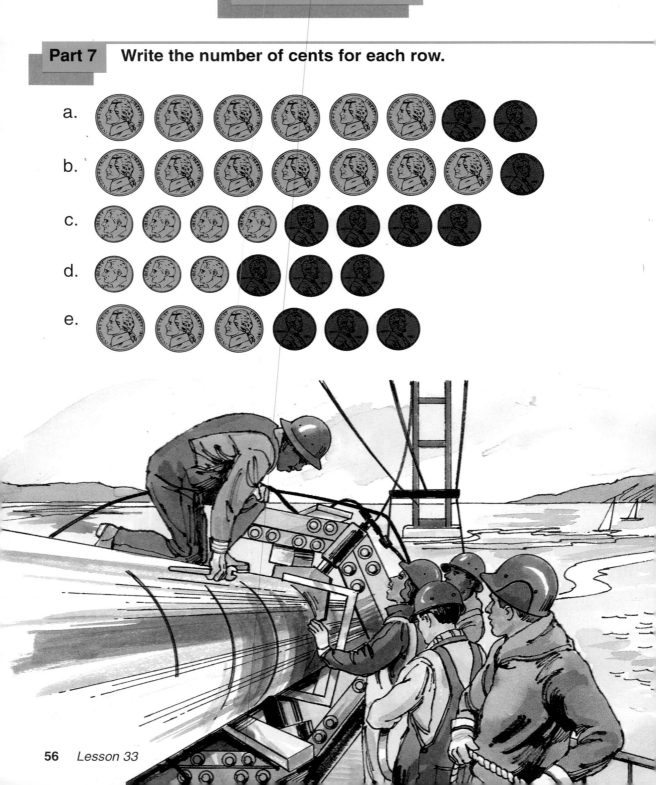

a.

b.

c.

d.

e.

Lesson 34

Part 1

For each problem, make a number family.
Then write the addition problem or subtraction problem
and figure out what number the letter is.

a. 194 is (329) less than J. What number is J?

b. R is (261) smaller than 684. What number is R?

c. 961 is (691) bigger than M. What number is M?

d. B is (365) more than 476. What number is B?

Part 2

Write the two multiplication facts for each number family.
Start the first fact with the first small number.

a. 5 ⌐—— 9 ——→ 45   b. 7 ⌐—— 10 ——→ 70

Part 3

First draw rectangles for each description.
Then write the numbers to show how high and how wide
each rectangle is.
Next to each rectangle write the area problem and the
answer.

a. The rectangle is 9 feet high and 5 feet wide.

b. The rectangle is 10 feet high and 2 feet wide.

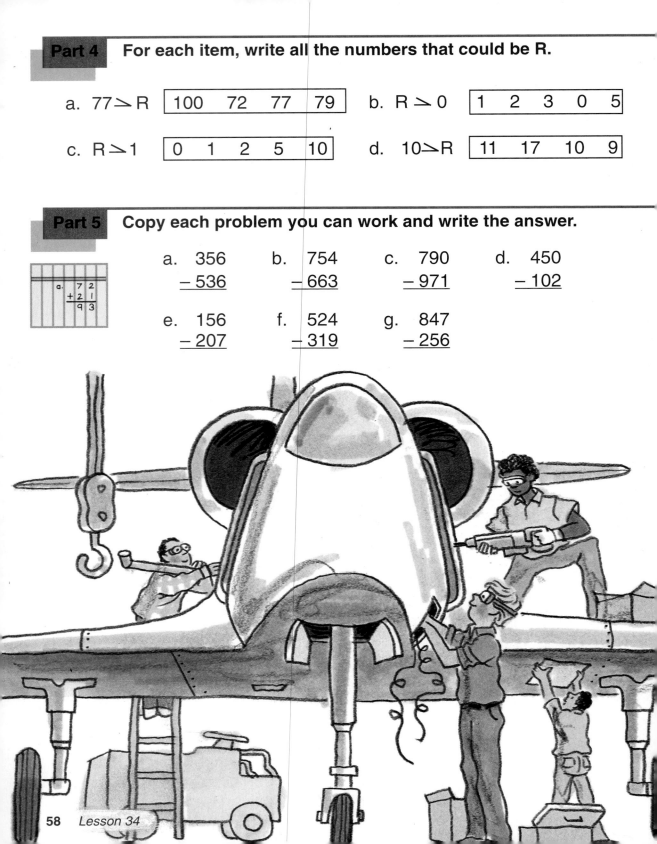

Part 4   For each item, write all the numbers that could be R.

a. $77 \succeq R$ | 100   72   77   79 |

b. $R \succeq 0$ | 1   2   3   0   5 |

c. $R \succeq 1$ | 0   1   2   5   10 |

d. $10 \succeq R$ | 11   17   10   9 |

Part 5   Copy each problem you can work and write the answer.

a.	356	b.	754	c.	790	d.	450
	$-\,536$		$-\,663$		$-\,971$		$-\,102$

e.	156	f.	524	g.	847
	$-\,207$		$-\,319$		$-\,256$

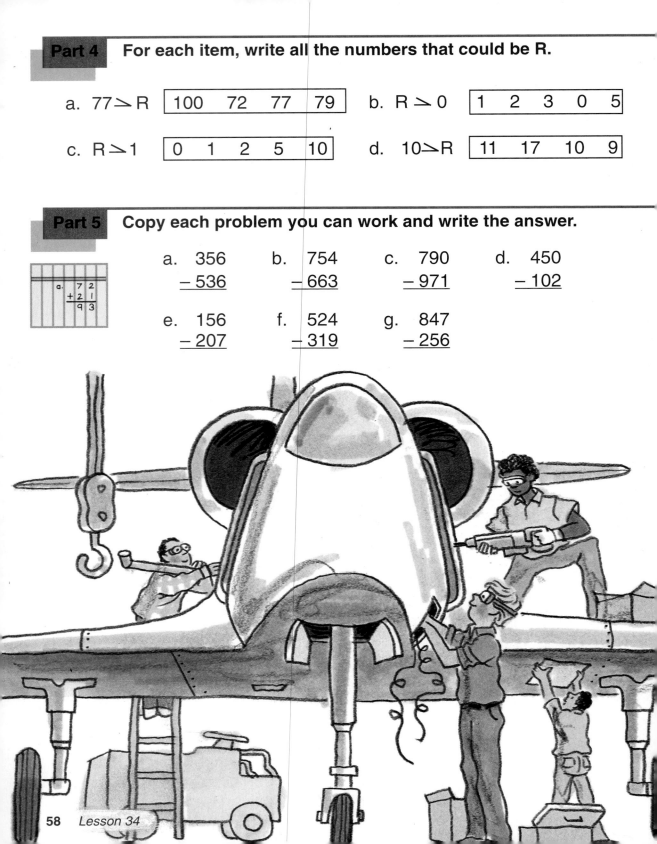

Lesson 35

Part 1

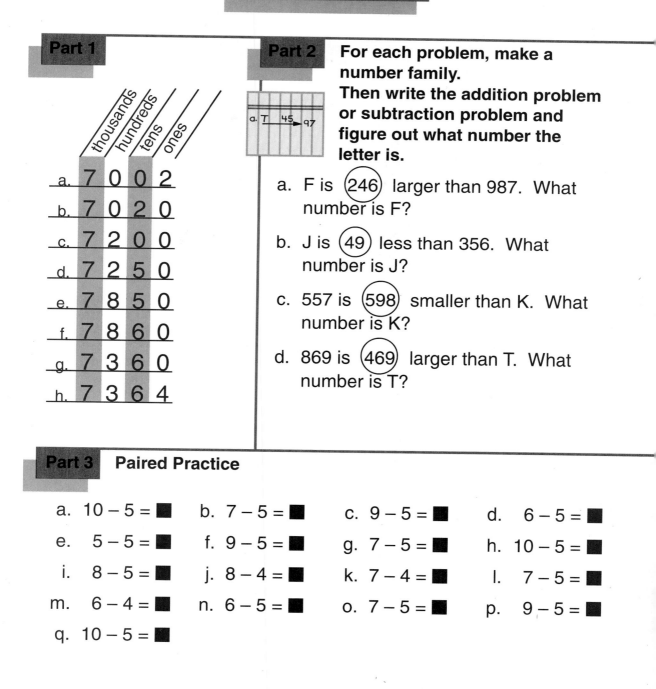

	thousands	hundreds	tens	ones
a.	7	0	0	2
b.	7	0	2	0
c.	7	2	0	0
d.	7	2	5	0
e.	7	8	5	0
f.	7	8	6	0
g.	7	3	6	0
h.	7	3	6	4

Part 2 For each problem, make a number family.
Then write the addition problem or subtraction problem and figure out what number the letter is.

a. F is ⟨246⟩ larger than 987. What number is F?

b. J is ⟨49⟩ less than 356. What number is J?

c. 557 is ⟨598⟩ smaller than K. What number is K?

d. 869 is ⟨469⟩ larger than T. What number is T?

Part 3   Paired Practice

a. $10 - 5 = \blacksquare$ b. $7 - 5 = \blacksquare$ c. $9 - 5 = \blacksquare$ d. $6 - 5 = \blacksquare$

e. $5 - 5 = \blacksquare$ f. $9 - 5 = \blacksquare$ g. $7 - 5 = \blacksquare$ h. $10 - 5 = \blacksquare$

i. $8 - 5 = \blacksquare$ j. $8 - 4 = \blacksquare$ k. $7 - 4 = \blacksquare$ l. $7 - 5 = \blacksquare$

m. $6 - 4 = \blacksquare$ n. $6 - 5 = \blacksquare$ o. $7 - 5 = \blacksquare$ p. $9 - 5 = \blacksquare$

q. $10 - 5 = \blacksquare$

- These are word problems that involve area.
- For each problem, you're going to draw a diagram of the problem.
- Then you'll write the multiplication problem and the answer.

a. A floor is the shape of a rectangle. It is 10 feet long and 8 feet wide. What is the area of the floor?

b. A large field is the shape of a rectangle. It is 3 miles long and 5 miles wide. What is the area of the field?

Part 5 **Copy each problem you can work and write the answer.**

a.	7	2
+	2	1
	9	3

a. 516
  − 409

b. 854
  − 264

c. 265
  − 584

d. 176
  − 348

e. 723
  − 404

f. 604
  − 721

g. 508
  − 347

Lesson 36

Part 1

a. 3 3 2 5   d. 7 5 0 0

b. 3 5 2 5   e. 7 5 2 0

c. 3 0 2 5   f. 7 5 2 5

Part 2

- The number that comes just before the words **more** or **less** is a small number.

- That's the number you write at the beginning of the number family.

- Here's a problem: D is 56 more than 221. $\xrightarrow{\quad 56 \quad\quad 221 \quad}$ D

- The number just before the word **more** is 56. That's a small number.

- Here's another problem: 562 is 89 less than D. $\xrightarrow{\quad 89 \quad\quad 562 \quad}$ D

- The number just before the word **less** is 89. That's a small number.

For each problem make a number family.
Then add or subtract to find out what number the letter is.

a. J is 56 less than 78. What number is J?

b. 400 is 90 more than T. What number is T?

c. 560 is 600 less than H. What number is H?

d. T is 380 more than 699. What number is T?

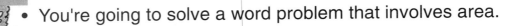

Part 3

- You're going to solve a word problem that involves area.
- First you'll draw a diagram of the rectangle.
- Then you'll write the multiplication problem and the whole answer.

a. A garden is the shape of a rectangle. It is 9 feet long and 6 feet wide. What is the area of the garden?

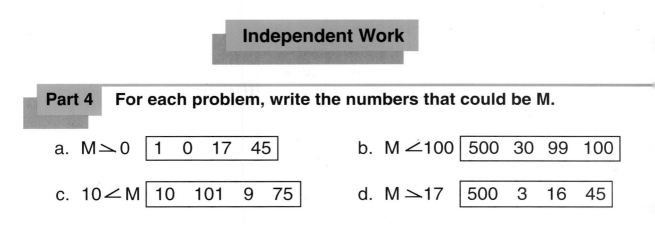

Independent Work

Part 4 For each problem, write the numbers that could be M.

a. M ＞ 0 | 1   0   17   45 |

b. M ＜ 100 | 500   30   99   100 |

c. 10 ＜ M | 10   101   9   75 |

d. M ＞ 17 | 500   3   16   45 |

Part 5 For each item, write the column problem and figure out the answer.

a. Jasmin had 147 snails. She found 23 more snails under a rock. How many snails did she end up with?

b. Carl had 247 balloons. 136 ballons popped. How many balloons did Carl end up with?

Part 6 Copy each problem you can work and write the answer.

	a. 497	b. 200	c. 605	d. 563	e. 573
	− 200	− 497	− 325	− 119	− 482

Lesson 37

Part 1

a. 3700

b. 3750

c. 3050

d. 3005

e. 8120

f. 4135

g. 6105

h. 2060

Part 2

Make a number family for each problem. Then add or subtract to find what number the letter is.

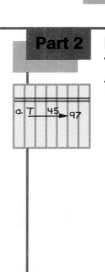

a. J is 13 more than 97. What number is J?

b. 58 is 290 less than P. What number is P?

c. 795 is 394 more than W. What number is W?

d. R is 200 less than 581. What number is R?

Part 3

Write the time shown on each clock.

a.

b.

c.
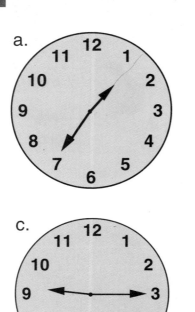

d.
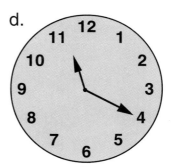

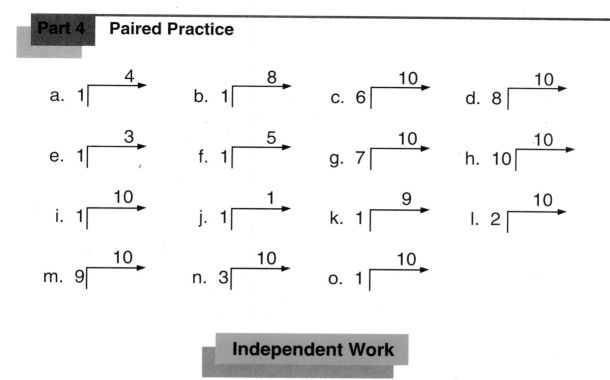

a. 1 → 4
b. 1 → 8
c. 6 → 10
d. 8 → 10

e. 1 → 3
f. 1 → 5
g. 7 → 10
h. 10 → 10

i. 1 → 10
j. 1 → 1
k. 1 → 9
l. 2 → 10

m. 9 → 10
n. 3 → 10
o. 1 → 10

Independent Work

Part 5   **Write the column problem and figure out the answer.**

a. There were 637 fleas on a dog. 74 more fleas jumped onto the dog. How many fleas are on the dog now?

b. Sue had 496 raisins. Her kids ate 247 raisins. How many raisins did Sue end up with?

Part 6   **Copy each problem and write the answer.**

a. 37
  + 88

b. 567
  + 949

c. 28
    11
  + 56

d. 49
     5
  + 21

For each problem, write the area problem and the answer.

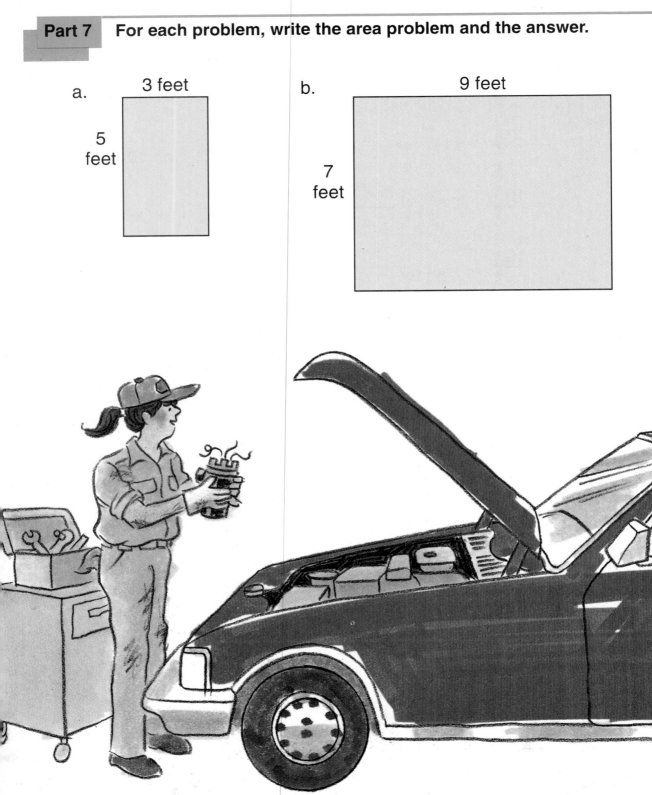

a.

3 feet

5
feet

b.

9 feet

7
feet

Lesson 38

Part 1   Use lined paper. Write the numeral for each item.

a. Seven thousand four hundred three.

b. Seven thousand forty.

c. Seven thousand four.

d. Seven thousand four hundred.

e. Four thousand twenty.

f. Four thousand two.

Part 2   Make the number family.
Then write the addition problem or subtraction problem.
Then answer the question the problem asks about.

a. S is 15 more than B. B is 77. What number is S?

b. F is 66 less than T. F is 399. What number is T?

c. J is 185 more than M. J is 276. What number is M?

d. L is 207 less than R. R is 288. What number is L?

Part 3   Write the addition problem for each item.

a. You have 2 threes.

b. You have 2 eights.

c. You have 2 tens.

d. You have 3 tens.

e. You have 4 tens.

f. You have 2 sevens.

Part 4 A person is not buying all four items with price tags.
Read each problem to see what the person buys.
Add up only those amounts.

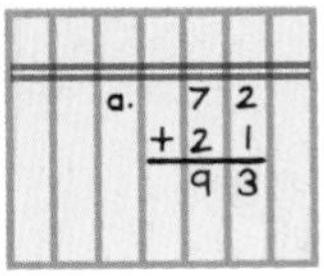

a.
	7	2
+	2	1
	9	3

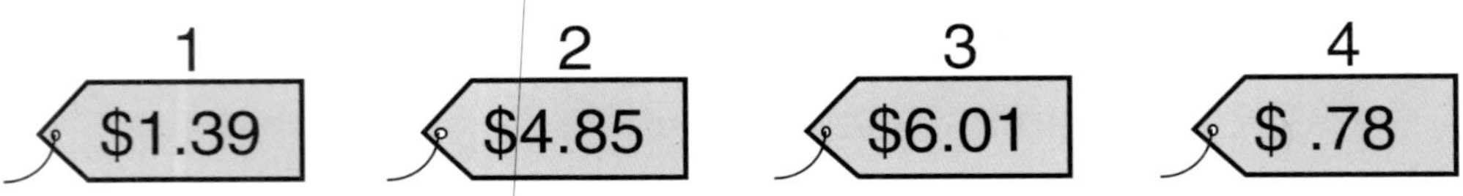

1 $1.39

2 $4.85

3 $6.01

4 $ .78

a. A person buys items 1, 2 and 3. How much does the person spend?

b. A person buys items 2 and 4. How much does the person spend?

c. A person buys items 1, 3 and 4. How much does the person spend?

Part 5 Write the times shown on each clock.

a.

b.

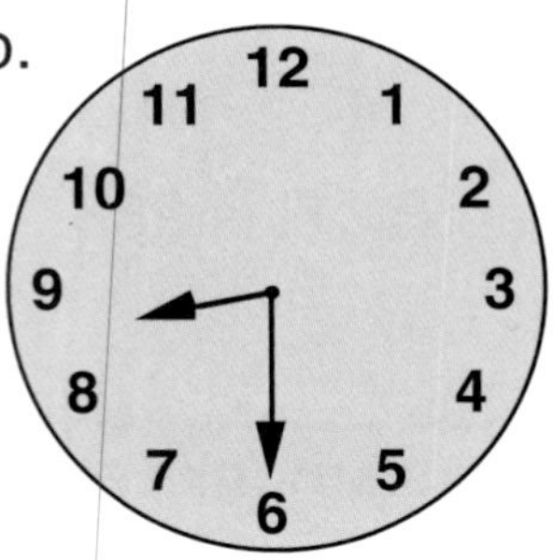

c.

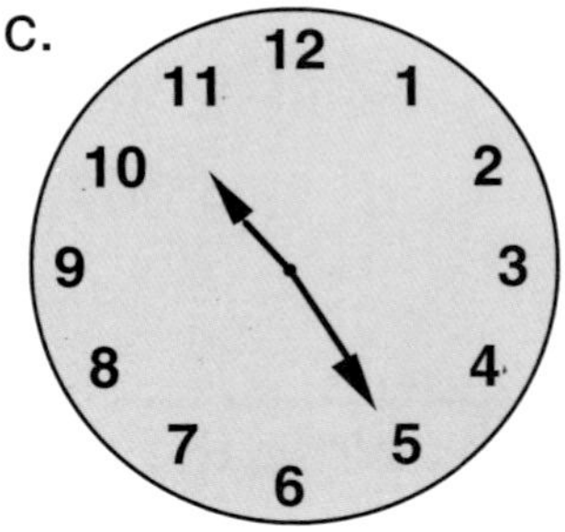

Part 6 For each item, write the numbers that could be H.

a. H < 10 | 10   0   15   101 |

b. H > 0 | 10   0   15   101 |

c. 10 > H | 0   3   10   101 |

d. 100 < H | 10   3000   970   100 |

For each item, write the area problem and the answer.

a.
7 miles

2
miles

b.
9 miles

7
miles

c.
3 miles

10
miles

d.
5 miles

5
miles

Part 1

- Here's how to work these problems:

1. Read the first sentence and make the number family with two letters and a number.

2. Then read the next sentence. That sentence gives a number for one of the letters. Cross out the letter and write the number.

3. Write the number problem and figure out the answer.

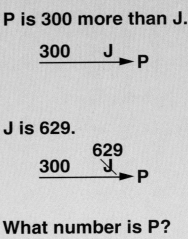

P is 300 more than J.

J is 629.

What number is P?

$$\begin{array}{r} 300 \\ + 629 \\ \hline 929 \end{array}$$

Make each number family with two letters. Replace one of the letters with a number. Then figure out the missing number.

a. R is 250 more than P. R is 881. What number is P?

b. J is 596 more than M. M is 387. What number is J?

c. K is 497 less than T. T is 797. What number is K?

d. W is 367 less than M. W is 928. What number is M?

Part 2   For each item, write the addition problem and the answer.

a. You have 3 tens.

b. You have 2 fives.

c. You have 2 tens.

d. You have 4 tens.

e. You have 2 threes.

f. You have 2 nines.

Part 3   Write the numeral for each item.

a. Three thousand twenty

b. Five thousand seventy

c. Five thousand seven hundred

d. Five thousand seven hundred fifty

e. Three thousand two

Part 4   Write the letter shown on the grid for each item.

> The X arrow is along the bottom. The Y arrow is up. To find the point, you first go along the X arrow. Then up for Y.

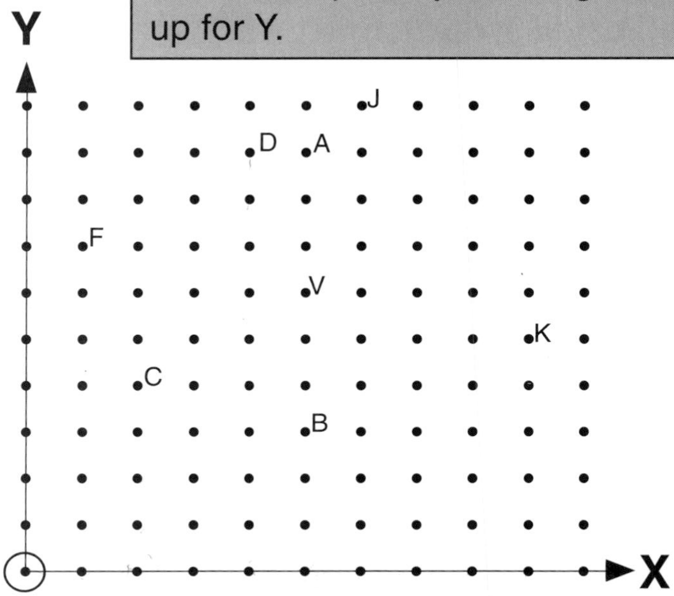

a. Letter ■
   (X = 6, Y = 10)

b. Letter ■
   (X = 4, Y = 9)

c. Letter ■
   (X = 2, Y = 4)

d. Letter ■
   (X = 9, Y = 5)

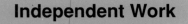
Part 5 **For each item draw a diagram.**
Then write the multiplication problem and the whole answer.

a. A field is the shape of a rectangle. The field is 9 miles long and 3 miles wide. What is the area of the field?

b. A floor tile is the shape of a rectangle. The tile is 6 inches long and 5 inches wide. What is the area of the floor tile?

Part 6 **Write the column problem and figure out the answer.**

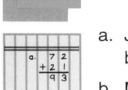

a. Joe had 634 bottles. He broke 144 bottles. How many bottles did Joe end up with?

b. Mary had 512 pins. She used 109 pins. How many pins did Mary end up with?

Part 7 **Write the number of cents in each row.**

a.

b.

c.

Write the area problem and the answer for each rectangle.

a.

8

5

b.

2

6

Lesson 40

Part 1   Write the letter shown on the grid for each item.

a. Letter ■
   (X = 7, Y = 1)

b. Letter ■
   (X = 2, Y = 10)

c. Letter ■
   (X = 5, Y = 5)

d. Letter ■
   (X = 9, Y = 9)

<u>M</u>ary has more fish than <u>H</u>enry.

- The first letter of each name is underlined—M and H.
- Mary has more, so Mary is the big number and Henry is a small number.
- Here's the number family with the letters for Mary and Henry:

$$\xrightarrow{\quad\quad\quad} \overset{H}{}\text{M}$$

<u>J</u>ane has fewer fish than <u>T</u>ed.

- Jane has fewer, so Jane is a small number. Ted is the big number.
- You write T for Ted and J for Jane.
- Here's the number family: $\xrightarrow{\quad\quad\quad} \overset{J}{}\text{T}$

For each sentence, write two letters in a number family. Write the first letter of each person's name.

a. <u>S</u>am is shorter than <u>J</u>im.

b. <u>A</u>nn has fewer coins than <u>T</u>racy.

c. <u>P</u>aul is taller than <u>G</u>inger.

d. <u>A</u>nn is older than <u>B</u>arbara.

e. <u>T</u>im is shorter than <u>W</u>endy.

f. <u>B</u>ob is younger than <u>T</u>im.

Part 3 **For each problem, make a number family. Replace one of the letters with a number. Then figure out the missing number.**

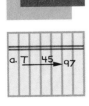

a. T is 480 less than B. B is 790. What number is T?

b. R is 386 less than M. R is 479. What number is M?

c. W is 227 less than P. W is 506. What number is P?

d. Y is 129 more than J. Y is 309. What number is J?

Write the subtraction problem and the answer.

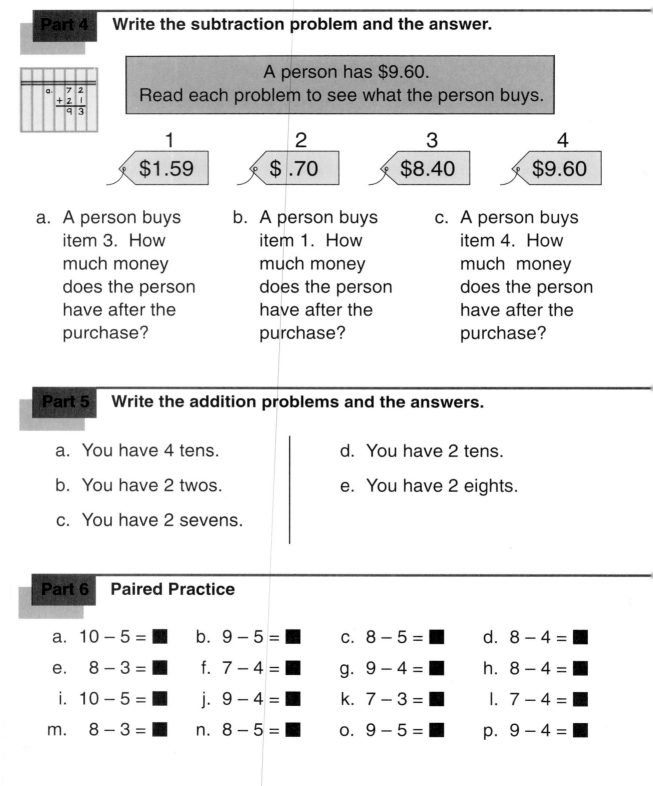

A person has $9.60.
Read each problem to see what the person buys.

1	2	3	4
$1.59	$ .70	$8.40	$9.60

a. A person buys item 3. How much money does the person have after the purchase?

b. A person buys item 1. How much money does the person have after the purchase?

c. A person buys item 4. How much money does the person have after the purchase?

Part 5 **Write the addition problems and the answers.**

a. You have 4 tens.

b. You have 2 twos.

c. You have 2 sevens.

d. You have 2 tens.

e. You have 2 eights.

Part 6 **Paired Practice**

a. $10 - 5 =$ ■   b. $9 - 5 =$ ■   c. $8 - 5 =$ ■   d. $8 - 4 =$ ■

e. $8 - 3 =$ ■   f. $7 - 4 =$ ■   g. $9 - 4 =$ ■   h. $8 - 4 =$ ■

i. $10 - 5 =$ ■   j. $9 - 4 =$ ■   k. $7 - 3 =$ ■   l. $7 - 4 =$ ■

m. $8 - 3 =$ ■   n. $8 - 5 =$ ■   o. $9 - 5 =$ ■   p. $9 - 4 =$ ■

Part 7 Write the missing big number for each number family.

a. 1 ⌐ 5 →

b. 10 ⌐ 10 →

c. 4 ⌐ 10 →

d. 1 ⌐ 7 →

e. 10 ⌐ 3 →

f. 8 ⌐ 1 →

g. 1 ⌐ 9 →

h. 2 ⌐ 10 →

i. 7 ⌐ 10 →

j. 6 ⌐ 1 →

k. 10 ⌐ 8 →

l. 3 ⌐ 1 →

Part 8 Write the number of cents in each row.

a.

b.

c.

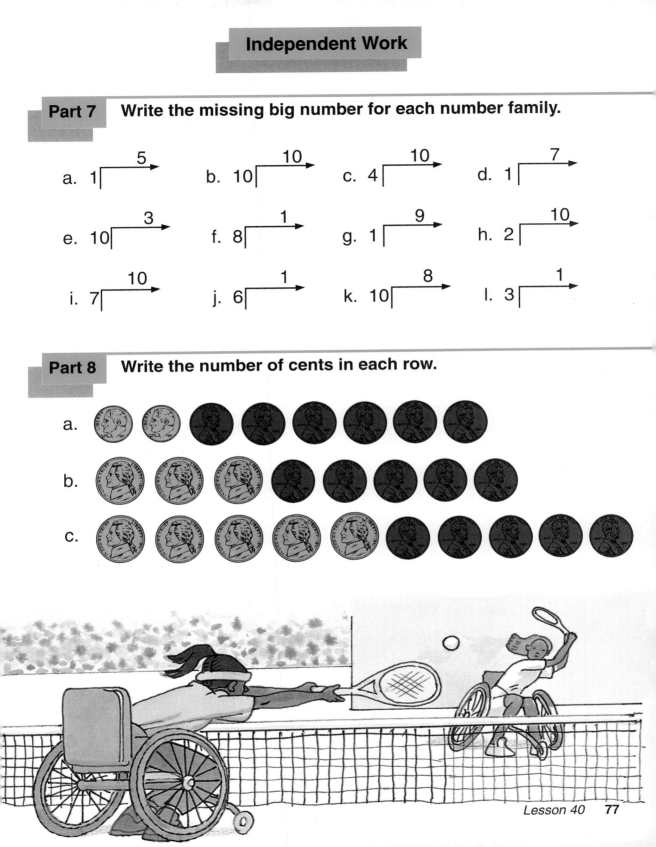

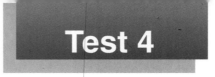

Test 4

Part 6   Draw the diagram and put the numbers in.
Then write the multiplication problem and the answer.

a. A floor is the shape of a rectangle. It is 10 feet long and 8 feet wide. What is the area of the floor?

b. A large field is the shape of a rectangle. It is 3 miles long and 5 miles wide. What is the area of the field?

Part 7   Make the number family. Then add or subtract to find out what number P is.

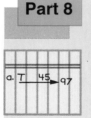

a. P is 26 less than 96. What number is P?

b. 59 is 31 more than P. What number is P?

c. 59 is 31 less than P. What number is P?

Part 8   For each problem, make a number family with two letters.
Replace one of the letters with a number.
Then figure out the other letter.

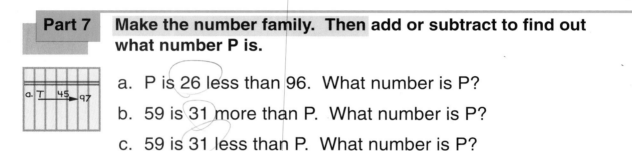

a. B is 250 more than T. B is 781. What number is T?

b. J is 394 less than K. J is 194. What number is K?

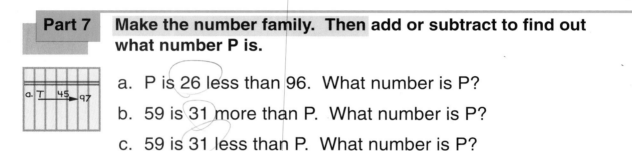

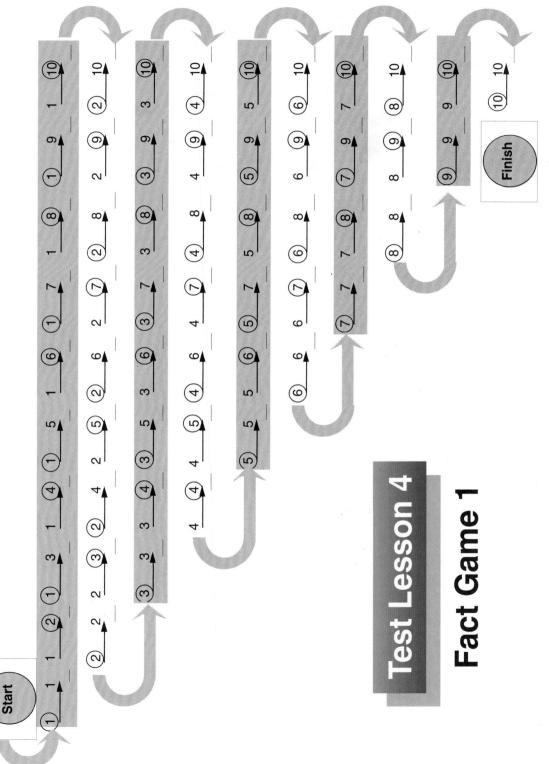

Test Lesson 4

Fact Game 1

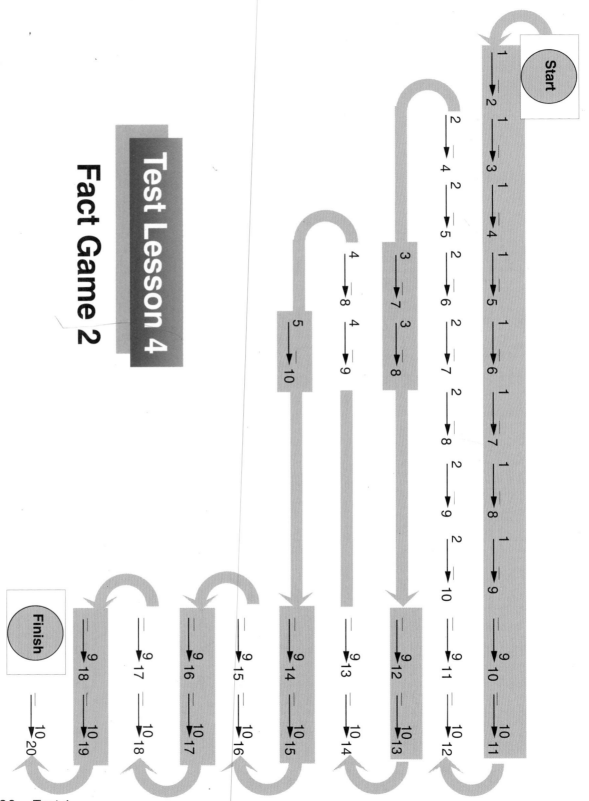

Test Lesson 4

Fact Game 2

Part 1 For each problem, make a number family with two letters.
Replace one of the letters with a number.
Then figure out the missing number.

a. M is 56 less than J. M is 250. What number is J?

b. R is 200 less than P. P is 700. What number is R?

c. T is 123 greater than P. T is 545. What number is P?

d. M is 72 greater than H. H is 672. What number is M?

Part 2

- When you make number families from word problems, you can use letters to stand for the names. You can write the first letter for each name in the problem.

- Here's a sentence: **Wendy has more seeds than Tom.**

- Here's the number family: $\xrightarrow[\hspace{3cm}]{\text{T}} \text{W}$

- Wendy is the W. She's the big number because she has more.

- Tom is the T. He's a small number in this problem.

Write the number family for each sentence.

a. Ann is older than Debby.

b. Doug runs less than Jerry.

c. Jan has more paper than Carol.

d. Tony is shorter than Jack.

e. Fran eats less than Don.

Part 3 Write both multiplication facts for each family.
Start the first fact with the **first** small number.

a. 1 ⌐——8——→ __   b. 6 ⌐——10——→ __

Part 4 For each problem, write the subtraction problem and figure
out the answer.

	7	2
a.	7	2
+ 2	1	
9	3	

1   $6.75   2 $2.09   3 $1.94

a. A man has $7.53. He buys
item 3. How much money
does he end up with?

b. A woman has $7.68. She
buys item 1. How much
money does she end up
with?

Part 5 Answer the questions about each numeral.

7316   599

a. How many digits are in 7316?

b. How many thousands?

c. How many tens?

d. How many ones?

e. How many digits are in 599?

f. How many hundreds?

g. How many tens?

h. How many ones?

Part 6 Write the time shown on each clock.

a.

b.

c.

d.

Lesson 42

Part 1

$$4 \qquad 8 \qquad 12 \qquad 16 \qquad 20 \longrightarrow$$

$$24 \qquad 28 \qquad 32 \qquad 36 \qquad 40 \longrightarrow$$

Part 2   Write the number family for each sentence.

a. Brian is younger than Debby.

b. Sid has less money than George.

c. Fran grew more trees than Jack did.

d. Tony weighs less than Jan.

e. Carol ran farther than Dorothy.

Part 3   For each problem, make a number family with two letters. Replace one of the letters with a number. Then figure out the missing number.

a. J is 200 less than M. J is 620. What number is M?

b. T is 18 less than R. R is 609. What number is T?

c. P is 301 greater than H. P is 322. What number is H?

d. Y is 39 more than V. V is 18. What number is Y?

- These are big thousands numerals. You can tell they are thousands because they have 3 zeros. But they're not hard to read.

- Here's how you do it: You read the number before the comma, then you say **thousand.**

a. 36,000   b. 27,000   c. 281,000   d. 15,000

- To find the X value for a point, go straight down from the point to the X arrow.

- To find the Y value, go straight across from the point to the Y arrow.

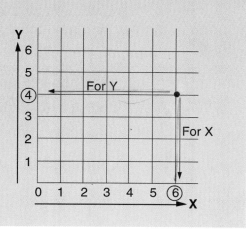

Independent Work

Part 6   **For each item, draw a diagram.**
Then write the multiplication problem and the whole answer.

a. A field is the shape of a rectangle. The field is 2 miles long and 7 miles wide. What is the area of the field?

b. A sheet of paper is 10 inches long and 6 inches wide. What is the area of the sheet of paper?

For each problem, write the subtraction problem and figure
out the answer.

1

$4.75

2

$1.19

a. A man has $11.62. He buys
item 2. How much money
does he have after the
purchase?

b. A woman has $6.84. She
buys item 1. How much
money does she have after
the purchase?

Part 8 Write the answers to the questions.

This table shows how many tons of corn each farmer
grew in 1982, 1983 and 1984.

	Jones	Smith	Green	Total
1982	7	9	10	26
1983	4	8	11	23
1984	8	3	0	11
Total	19	20	21	

a. Which farmer grew the most
corn?

b. The fewest tons of corn were
grown in which year?

c. Farmer Green did not grow
corn in which year?

d. Who grew the fewest tons of
corn?

Lesson 43

Part 1   Copy each problem and work it.

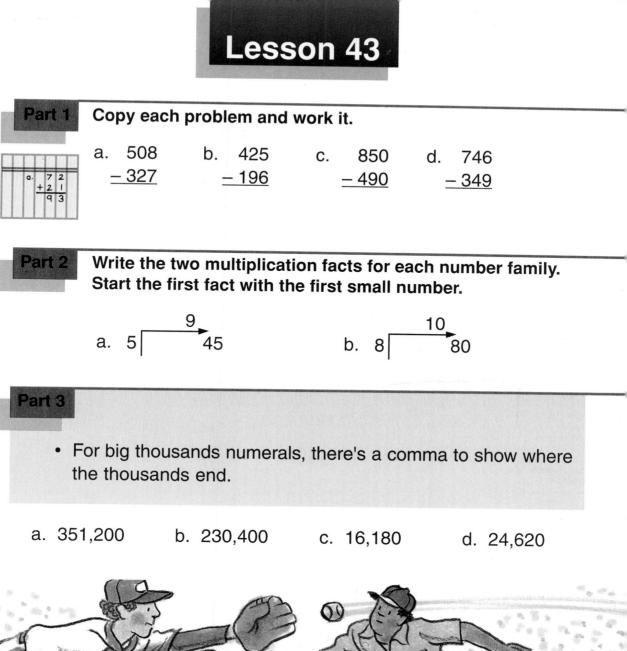

a.   508
   − 327

b.   425
   − 196

c.   850
   − 490

d.   746
   − 349

Part 2   Write the two multiplication facts for each number family. Start the first fact with the first small number.

a.   5 ⟶⁹ 45

b.   8 ⟶¹⁰ 80

Part 3

- For big thousands numerals, there's a comma to show where the thousands end.

a. 351,200   b. 230,400   c. 16,180   d. 24,620

- Here's how to work the problems:

1. Read the first sentence and make the number family with two letters and a number.

2. Then read the next sentence. That sentence gives a number for one of the letters. Cross out the letter and write the number.

3. Figure out the number for the other letter.

4. Write the number problem and the answer.

- **Fran was 14 years older than Ann.** $\underrightarrow{\quad 14 \qquad A \quad} F$

- **Ann was 13 years old.** $\underrightarrow{\quad 14 \qquad \overset{13}{\cancel{A}} \quad} F$

- **How many years old was Fran?** $\underrightarrow{\quad 14 \qquad \overset{13}{\cancel{A}} \quad} \overset{27}{\cancel{F}}$

$$\begin{array}{r} 1\ 4 \\ +\ 1\ 3 \\ \hline 2\ 7 \end{array}$$

Write the complete number family.
Then write the addition problem or subtraction problem and the answer.

a. Jane was ⑰ years younger than Bill. Bill was 56 years old. How many years old was Jane?

b. Sam was 52 years younger than Ginger. Sam was 31 years old. How many years old was Ginger?

c. Fran was 11 years older than Ron. Fran was 31 years old. How many years old was Ron?

d. Jan was 14 years younger than Al. Jan was 34 years old. How many years old was Al?

Part 5 Write the numbers that the letters could be.

a. K ⪈ 17 | 17 16 160 3000 | b. M ⪇ 1 | 1 1000 10 0 |

c. L ⪇ 1000 | 908 12 8000 101 | d. 15 ⪇ N | 18 14 140 15 |

Part 6 Write the subtraction problems and the answers.

```
   7 2
 + 2 1
 ─────
   9 3
```

1
$1.94

2
$2.48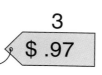

3
$ .97

a. A woman has $14.27. She buys item 2. How much money does she have after the purchase?

b. A woman has $14.27. She buys item 3. How much money does she have after the purchase?

Part 7 Copy and work each problem.

```
   7 2
 + 2 1
 ─────
   9 3
```

a. 518   b.   49   c.   29
     29   29   339
 +   3   12   +   13
 ─────   +   5
                  ─────
```

**Part 8**  Write the addition problems and the answers.

a. You had 2 sixes.

b. You have 3 tens.

c. You have 4 fives.

d. You have 3 twos.
```

Lesson 44

Part 1 Copy each problem and work it.

a.	8060	b.	3241	c.	7810	d.	6789
	− 2250		− 1331		− 1900		− 2890

Part 2

a. 27,100   b. 290,040   c. 37,004   d. 56,125   e. 300,001

Part 3

For each problem, make a number family with two letters. Replace one of the letters with a number. Then write the addition problem or subtraction problem and the answer.

a. Tim was 7 inches taller than Bill. Bill was 54 inches tall. How many inches tall was Tim?

b. Fran was 13 inches shorter than Ginger. Ginger was 54 inches tall. How many inches tall was Fran?

c. Debby was 21 inches taller than Billy. Billy was 37 inches tall. How many inches tall was Debby?

d. Reggie was 17 inches shorter than Greg. Greg was 78 inches tall. How many inches tall was Reggie?

Part 4 For each item, draw a diagram.
Then write the multiplication problem and the whole answer.

a. A park is the shape of a rectangle. The park is 4 miles long and 2 miles wide. What is the area of the park?

b. A room is the shape of a rectangle. The room is 10 feet long and 10 feet wide. What is the area of the room?

Part 5 Make a number family for each item.
The write the addition or subtraction problem and figure out what the letter is.

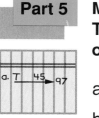

a. J is 144 less than 242. What number is J?

b. 916 is 247 more than M. What number is M?

c. 212 is 400 less than M. What number is M?

d. B is 57 less than 248. What number is B?

Part 6 Answer each question.

	big trucks	small trucks	total trucks
Maple Street	1	10	11
Pine Street	3	4	7
Total for both streets	4	14	

a. How many small trucks went down Maple Street?

b. How many small trucks went down both streets?

c. Were there more small trucks or big trucks on both streets?

d. How many big trucks went down Pine Street?

Part 7 For each item, copy the X and Y values. Then write the letter shown on the grid.

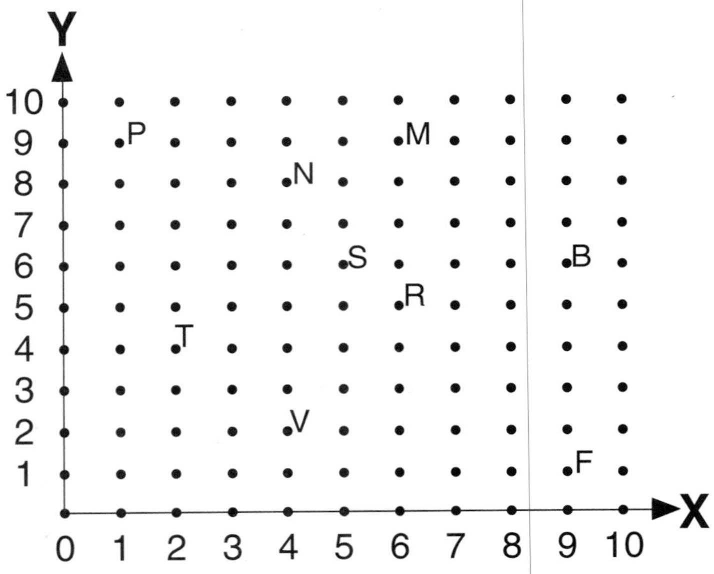

a. Letter ■ (X = 1, Y = 9)

b. Letter ■ (X = 5, Y = 6)

c. Letter ■ (X = 6, Y = 9)

d. Letter ■ (X = 4, Y = 2)

e. Letter ■ (X = 9, Y = 1)

Lesson 45

Part 1

Write the problem and the answer for each number family.

a. $\xrightarrow{\quad 4 \quad \blacksquare \quad} 12$ b. $\xrightarrow{\quad \blacksquare \quad 5 \quad} 8$ c. $\xrightarrow{\quad 3 \quad 10 \quad} \blacksquare$ d. $\xrightarrow{\quad \blacksquare \quad 9 \quad} 16$

e. $\xrightarrow{\quad 7 \quad 8 \quad} \blacksquare$ f. $\xrightarrow{\quad 4 \quad \blacksquare \quad} 10$ g. $\xrightarrow{\quad \blacksquare \quad 5 \quad} 7$ h. $\xrightarrow{\quad 5 \quad 9 \quad} \blacksquare$

Part 2

   a. 400,000   b. 26,500   c. 41,210   d. 38,005   e. 42,050

Part 3

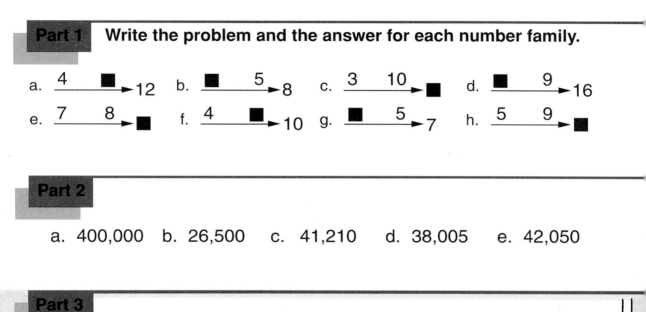

- These are word problems that tell about persons who are heavier and lighter.

- The person who is heavier is the **big number.**

- And the number just before the word **heavier** or **lighter** is **a small number.**

- Here's a sentence: **Tim is 13 pounds lighter than Jan.**

- Tim is lighter, so T is a small number. 13 is the other small number.

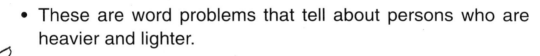

$\xrightarrow{\quad 13 \quad T \quad} J$

**Write the complete number family for each problem.
Replace one of the letters with a number.
Then write the addition problem or subtraction problem
and the answer.**

a. Hank was 40 pounds lighter than Bob. Hank weighed 130
   pounds. How many pounds was Bob?

b. Tina was 11 pounds heavier than Debby. Debby weighed 98
   pounds. How many pounds was Tina?

c. Ginger was 23 pounds lighter than Vern. Vern weighed 153
   pounds. How many pounds was Ginger?

d. Don was 36 pounds heavier than Kay. Don weighed 148
   pounds. How many pounds was Kay?

Independent Work

Part 4 **For each problem, write the subtraction problem and figure
out the answer.**

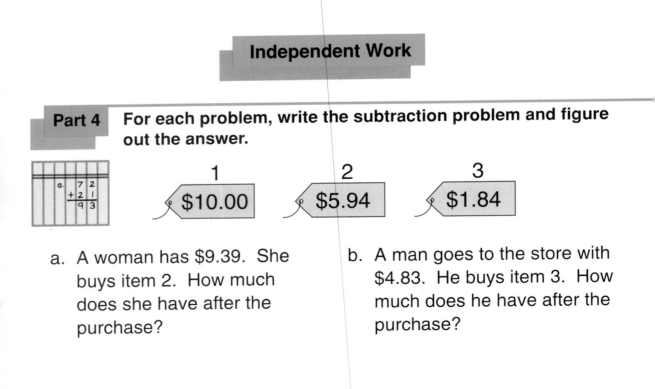

1 $10.00

2 $5.94

3 $1.84

a. A woman has $9.39. She
   buys item 2. How much
   does she have after the
   purchase?

b. A man goes to the store with
   $4.83. He buys item 3. How
   much does he have after the
   purchase?

Part 5 Write what X equals and what Y equals for each letter on the grid.

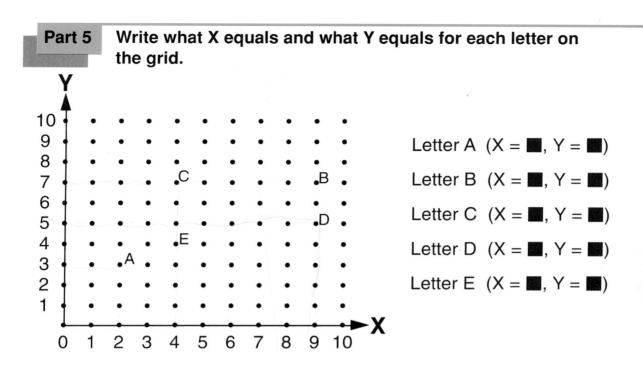

Letter A (X = ■, Y = ■)

Letter B (X = ■, Y = ■)

Letter C (X = ■, Y = ■)

Letter D (X = ■, Y = ■)

Letter E (X = ■, Y = ■)

Part 6 Write the column problem and figure out the answer.

a. Joe had 187 comic books. He gave away 168 comic books. How many comic books did Joe end up with?

b. Joan had 459 raisins. She bought 141 more raisins. How many raisins did she end up with?

Lesson 46

Part 1

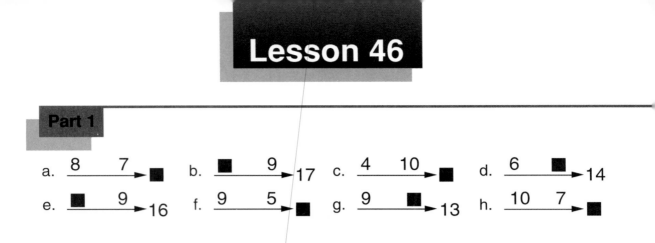

a. 8 → 7 → ■   b. ■ → 9 → 17   c. 4 → 10 → ■   d. 6 → ■ → 14

e. ■ → 9 → 16   f. 9 → 5 → ■   g. 9 → ■ → 13   h. 10 → 7 → ■

Part 2

**Make a number family for each problem.
The write the addition or subtraction problem and figure
out the answer.**

a. George was 17 years older than Janice. Janice was 38
years old. How old was George?

b. Alice weighed 22 pounds less than Greg. Greg weighed
131 pounds. How many pounds did Alice weigh?

c. Fran was 14 inches shorter than Ginger. Fran was 49
inches tall. How tall was Ginger?

d. Jerry was 31 years younger than his dad. His dad was
40 years old. How old was Jerry?

Part 3

a. 700,000   b. 21,000   c. 56,500   d. 91,025   e. 86,250

Part 4 For each problem, make a number family, write the column problem and figure out the answer.

a. L is 450 less than R. L is 69. What is R?

b. G is 39 more than J. G is 148. What is J?

Part 5 Copy each problem and figure out the answer.

a.　92　　b. 7637　　c.　996　　d.　436　　e. 5627
　　463　　　− 534　　　− 457　　　　9　　　− 249
　 + 109　　　　　　　　　　　　　+ 385

Part 6 Write what X equals and what Y equals for each letter on the grid.

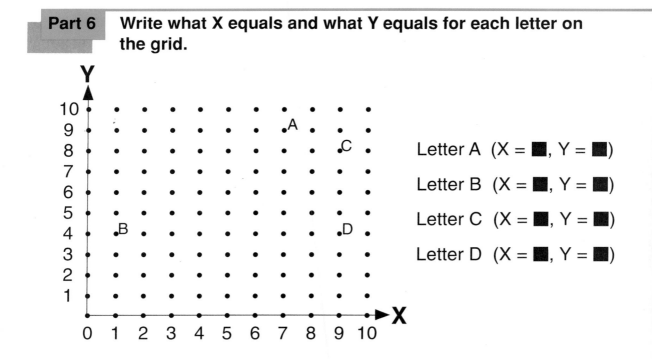

Letter A (X = ■, Y = ■)

Letter B (X = ■, Y = ■)

Letter C (X = ■, Y = ■)

Letter D (X = ■, Y = ■)

Part 7 Write the cents in each row.

a.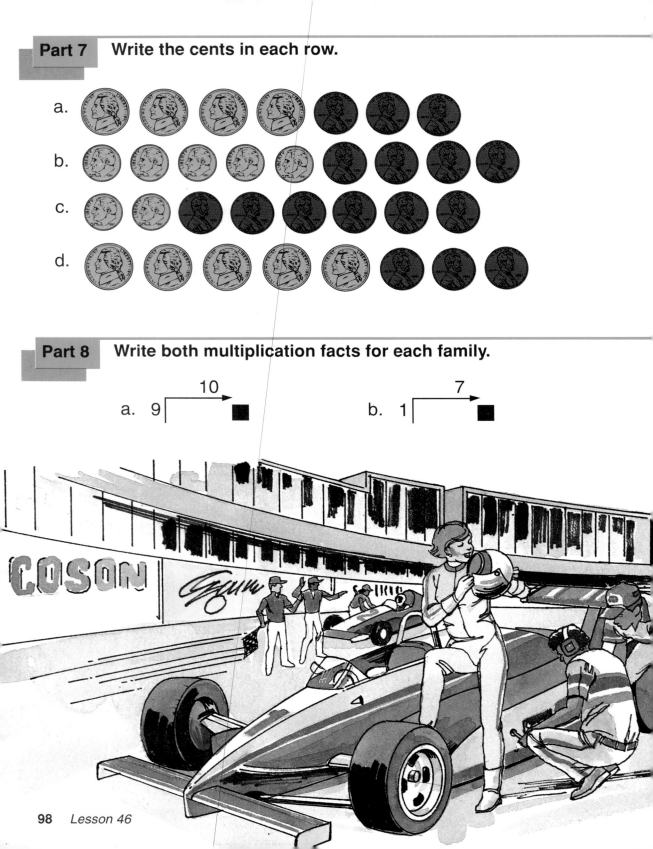

b.

c.

d.

Part 8 Write both multiplication facts for each family.

a. $9 \overset{10}{\big|\quad\rightarrow} \blacksquare$

b. $1 \overset{7}{\big|\quad\rightarrow} \blacksquare$

Lesson 47

Part 1

a. $\underrightarrow{\blacksquare \quad 49} 76$ b. $\underrightarrow{\blacksquare \quad 5} 15$ c. $\underrightarrow{28 \quad 172} \blacksquare$

d. $\underrightarrow{8 \quad \blacksquare} 9$ e. $\underrightarrow{156 \quad 28} \blacksquare$ f. $\underrightarrow{19 \quad \blacksquare} 280$

Part 2

**Make a number family for each problem.
Then write the addition or subtraction problem and figure
out the answer.**

a. Janice had 31 more bottle caps than Henry had. Janice
had 609 bottle caps. How many bottle caps did Henry
have?

b. Ted had 39 fewer marbles than Ginger had. Ginger had
158 marbles. How many marbles did Ted have?

c. Brian had 41 fewer cards than Tom had. Brian had 403
cards. How many cards did Tom have?

d. Fran had 81 more nails than Rita had. Rita had 490
nails. How many nails did Fran have?

Part 3

**Copy the table.
Write the column addition or subtraction problem for each
row. Then write the missing numbers in the table.**

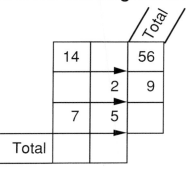

			Total
14			56
	2	9	
7	5		
Total			

Part 4 Write the answers to the questions.

The table shows the number of trucks and cars that went down Elm Street and Oak Street.

	Trucks	Cars	Total vehicles
Elm Street	9	4	13
Oak Street	12	2	14
Total for both streets	21	6	

a. How many trucks went down Oak Street?

b. How many total cars went down both the streets?

c. How many trucks went down Elm Street?

d. Were there more vehicles on Elm Street or Oak Street?

Part 5 For each item, draw a diagram.
Then write the multiplication problem and the whole answer.

a. A field is the shape of a rectangle. The field is 3 miles wide and 5 miles long. What is the area of the field?

b. A wall is the shape of a rectangle. The wall is 9 feet high and 4 feet wide. What is the area of the wall?

For each problem, write the numbers in the box that could be J.

a. J∠199 | 200 201 190 2000 | b. 79⟩J | 0 107 79 18 |

Copy each problem and figure out the answer.

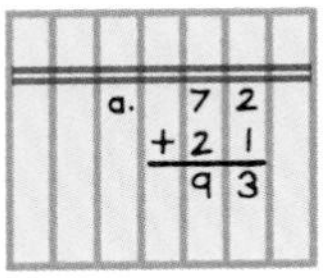

	7	2
+	2	1
	9	3

a. 825
  − 519

b. 666
  − 297

c. 485
  − 374

d. 39
      9
     12
  + 1

e. 56
     19
      5
  + 5

Lesson 48

Part 1

a. $\dfrac{9 \quad \blacksquare}{\longrightarrow 12}$ b. $\dfrac{406 \quad 120}{\longrightarrow \blacksquare}$ c. $\dfrac{14 \quad \blacksquare}{\longrightarrow 87}$

d. $\dfrac{\blacksquare \quad 2}{\longrightarrow 16}$ e. $\dfrac{20 \quad \blacksquare}{\longrightarrow 56}$ f. $\dfrac{490 \quad 12}{\longrightarrow \blacksquare}$

Part 2

**Make a number family for each problem.
Then write the addition or subtraction problem and figure
out the answer.**

a. Kathy picked 47 fewer apples than Debby. Debby
picked 246 apples. How many apples did Kathy pick?

b. George picked 103 more apples than Hank. Hank
picked 307 apples. How many apples did George pick?

c. Fran sold 29 more apples than Doug sold. Fran sold
318 apples. How many apples did Doug sell?

d. Brian sold 400 fewer apples than Ginger sold. Brian
sold 199 apples. How many apples did Ginger sell?

Part 3 Paired Practice

a. 2 x 7 = $\blacksquare$ b. 2 x 9 = $\blacksquare$ c. 2 x 4 = $\blacksquare$ d. 2 x 5 = $\blacksquare$

e. 2 x 10 = $\blacksquare$ f. 2 x 6 = $\blacksquare$ g. 4 x 2 = $\blacksquare$ h. 8 x 2 = $\blacksquare$

i. 9 x 2 = $\blacksquare$ j. 5 x 2 = $\blacksquare$ k. 3 x 2 = $\blacksquare$ l. 2 x 9 = $\blacksquare$

m. 6 x 2 = $\blacksquare$ n. 2 x 4 = $\blacksquare$ o. 7 x 2 = $\blacksquare$ p. 2 x 6 = $\blacksquare$

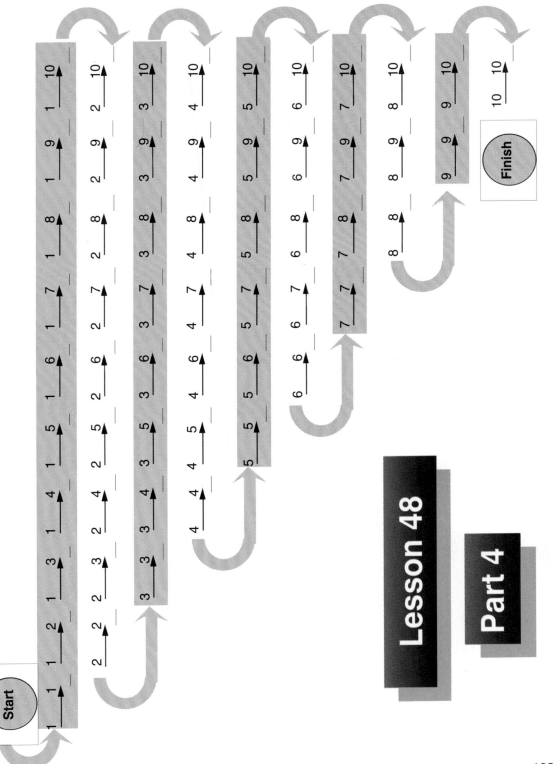

Part 5 For each item, draw a diagram.
Then write the multiplication problem and the whole answer.

a. A rectangular room is 7 feet wide and 9 feet long. What is the area of the room?

b. A rectangular piece of paper is 5 inches wide and 7 inches long. What is the area of the paper?

Part 6 Copy each problem and figure out the answer.

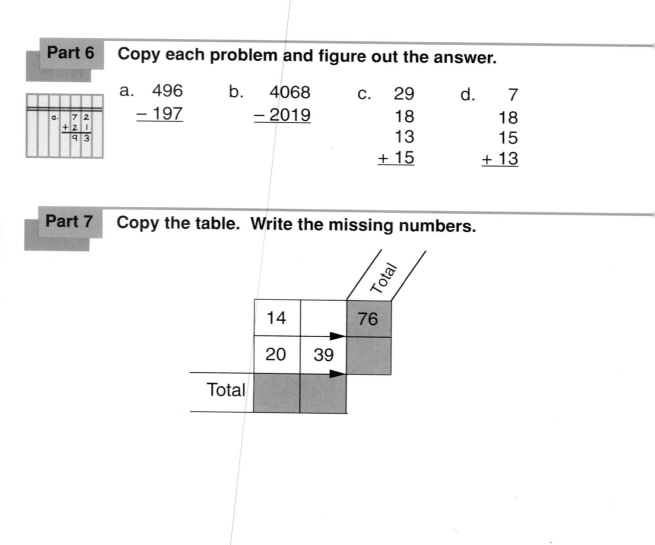

a.   496
  − 197

b.   4068
  − 2019

c.   29
  18
  13
  + 15

d.   7
  18
  15
  + 13

Part 7 Copy the table. Write the missing numbers.

		Total
14		76
20	39	
Total		

Lesson 49

Part 1 Work all the problems that have correct number families.

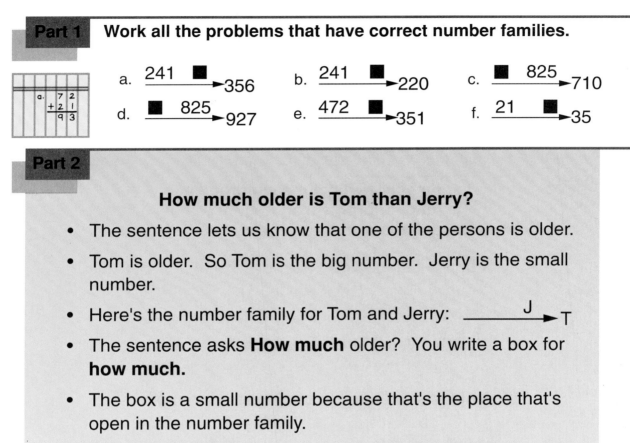

a. $\underset{\longrightarrow 356}{241 \quad \blacksquare}$

b. $\underset{\longrightarrow 220}{241 \quad \blacksquare}$

c. $\underset{\longrightarrow 710}{\blacksquare \quad 825}$

d. $\underset{\longrightarrow 927}{\blacksquare \quad 825}$

e. $\underset{\longrightarrow 351}{472 \quad \blacksquare}$

f. $\underset{\longrightarrow 35}{21 \quad \blacksquare}$

Part 2

How much older is Tom than Jerry?

- The sentence lets us know that one of the persons is older.

- Tom is older. So Tom is the big number. Jerry is the small number.

- Here's the number family for Tom and Jerry: $\underset{\longrightarrow T}{\overset{J}{\rule{2cm}{0.4pt}}}$

- The sentence asks **How much** older? You write a box for **how much.**

- The box is a small number because that's the place that's open in the number family.

- Here's the number number family for the sentence: **How much older is Tom than Jerry?** $\underset{\longrightarrow T}{\overset{J}{\square \rule{1.5cm}{0.4pt}}}$

Write the number family for each sentence.

a. How much shorter is Sally than Jane?

b. How much faster is Ginger than Dan?

c. How much younger is John than Kay?

d. How much taller is Barbara than Fran?

e. How much longer was the truck than the bus?

Part 3 Write the facts for each family in a column.

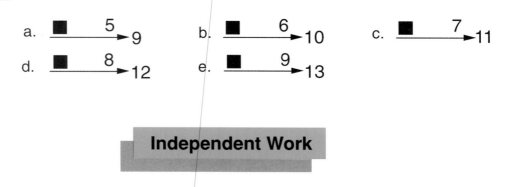

a. ■ ——5——→ 9   b. ■ ——6——→ 10   c. ■ ——7——→ 11

d. ■ ——8——→ 12   e. ■ ——9——→ 13

Independent Work

Part 4 Make the number family for each problem.
Write the number problem and figure out the answer.
Write the unit name in the answer.

a. A horse weighed 120 pounds less than a cow. The
horse weighed 730 pounds. How much did the cow
weigh?

b. A snake was 23 inches longer than a toad. The snake
was 32 inches long. How long was the toad?

c. A pile of coal was 17 feet higher than a pile of sand. The
pile of sand was 35 feet high. How high was the pile of
coal?

Part 5 Copy each problem and figure out the answer.

a. 7865   b. 6351   c. 572   d. 227
  − 264   − 1199   − 364   − 83

Part 6   Write the addition problem and the answer for each item.

a. You have 4 tens.

b. You have 2 eights.

c. You have 3 nines.

d. You have 5 twos.

Part 7   Find the area of each rectangle.

a.   5 inches

3 inches

b.   8 miles

4 miles

Part 8   Copy the table. Write the missing numbers.

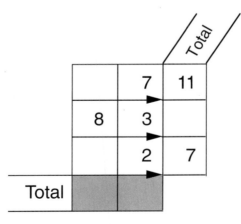

Lesson 50

Part 1

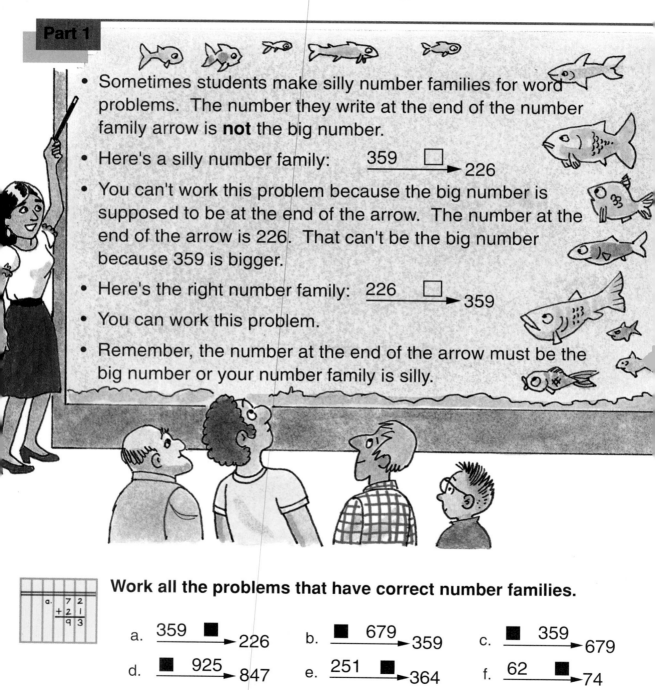

- Sometimes students make silly number families for word problems. The number they write at the end of the number family arrow is **not** the big number.

- Here's a silly number family: $\xrightarrow[\qquad]{359 \quad \square} 226$

- You can't work this problem because the big number is supposed to be at the end of the arrow. The number at the end of the arrow is 226. That can't be the big number because 359 is bigger.

- Here's the right number family: $\xrightarrow[\qquad]{226 \quad \square} 359$

- You can work this problem.

- Remember, the number at the end of the arrow must be the big number or your number family is silly.

Work all the problems that have correct number families.

a.
```
  7 2
+ 2 1
  9 3
```

a. $\xrightarrow[\qquad]{359 \quad \blacksquare} 226$

b. $\xrightarrow[\qquad]{\blacksquare \quad 679} 359$

c. $\xrightarrow[\qquad]{\blacksquare \quad 359} 679$

d. $\xrightarrow[\qquad]{\blacksquare \quad 925} 847$

e. $\xrightarrow[\qquad]{251 \quad \blacksquare} 364$

f. $\xrightarrow[\qquad]{62 \quad \blacksquare} 74$

Part 2   **Make a number family for each sentence.**

a. How much older is Jan than Tina?

b. How much slower is Don than Frank?

c. How much heavier is Sam than Dan?

d. How much taller is Rico than Jose?

Part 3

a.

■ ■ →7

■ ■ →7

■ ■ →7

b.

■ ■ →9

■ ■ →9

■ ■ →9

■ ■ →9

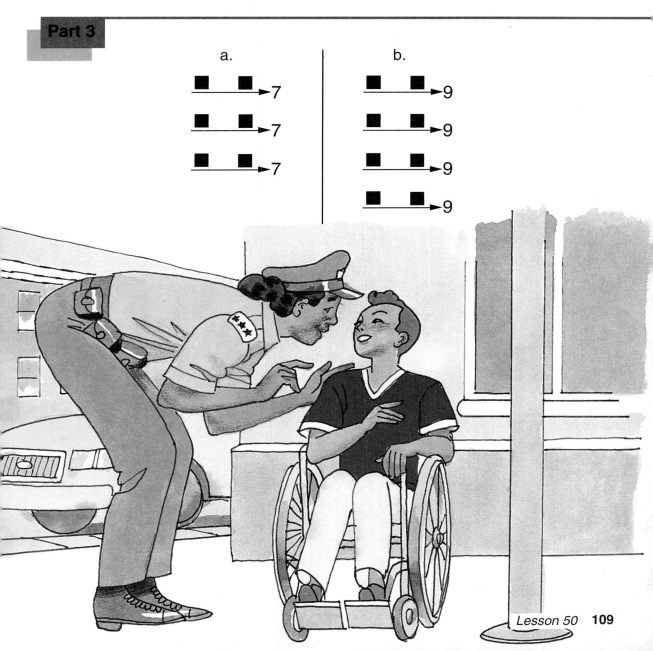

Part 4 For each item, write the numbers that could be T.

a. $32 \searrow T$ | 0   1000   32   31 | b. $T \angle 100$ | 10   99   0   100 |

Part 5 For each problem, write a number family and figure out the answer.

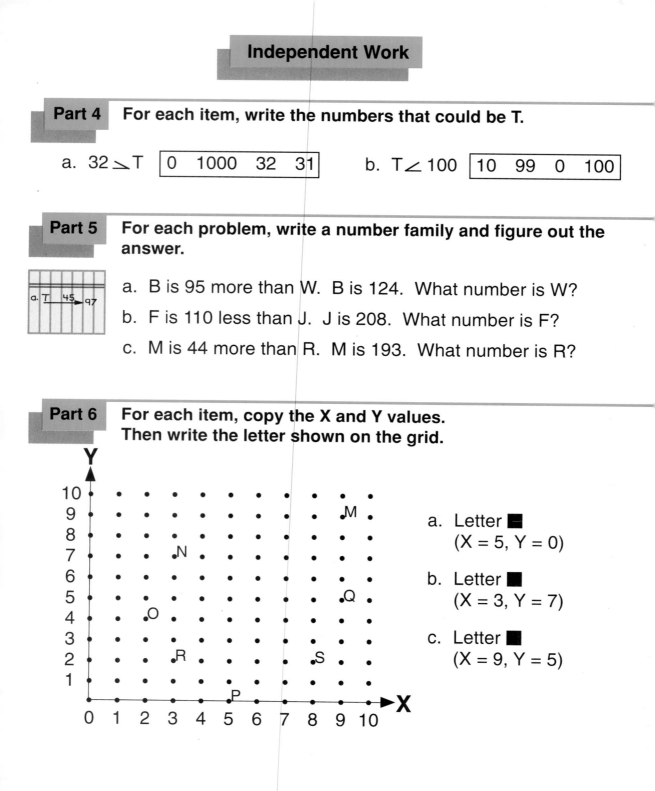

a. B is 95 more than W. B is 124. What number is W?

b. F is 110 less than J. J is 208. What number is F?

c. M is 44 more than R. M is 193. What number is R?

Part 6 For each item, copy the X and Y values.
Then write the letter shown on the grid.

a. Letter ■
    (X = 5, Y = 0)

b. Letter ■
    (X = 3, Y = 7)

c. Letter ■
    (X = 9, Y = 5)

Part 7 For each problem, make a number family.
Then write the number problem and the answer.
Write the unit name in the answer.

a. Tim is 24 years younger than his dad. His dad is 43 years old. How old is Tim?

b. A snake is 13 inches longer than a worm. The snake is 25 inches long. How long is the worm?

c. An elephant weighs 350 pounds less than a truck. The elephant weighs 7380 pounds. How much does the truck weigh?

Part 8 Write the answers to the questions.

The table shows the number of brown birds and green birds that were seen on Friday and Saturday.

	Friday	Saturday	Total for both days
Brown birds	12	18	30
Green birds	12	16	28
Total birds	24	34	

a. How many green birds were seen on Saturday?

b. How many total birds were seen on Friday?

c. Were more birds seen on Friday or on Saturday?

d. How many green birds were seen on Friday?

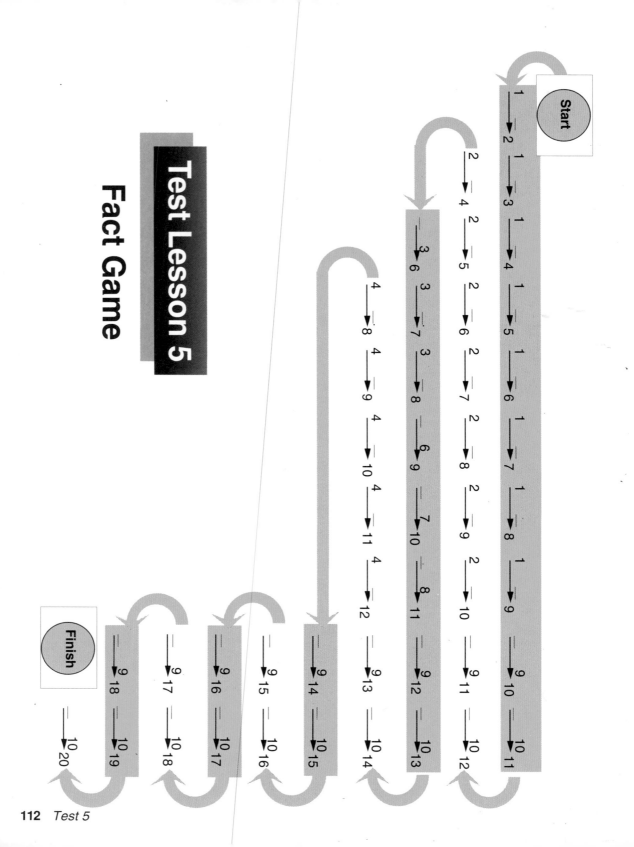

Test Lesson 5

Fact Game

Start

Lesson 51

Part 1

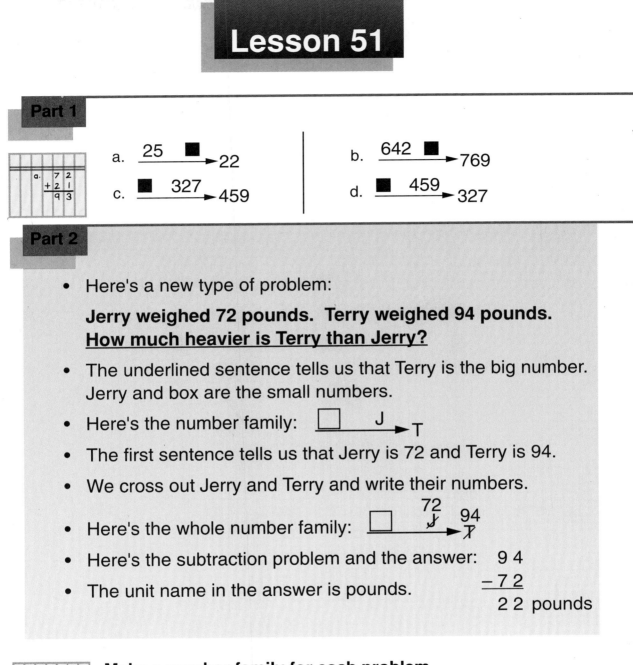

a. $\underline{25 \quad \blacksquare} \rightarrow 22$

b. $\underline{642 \quad \blacksquare} \rightarrow 769$

c. $\underline{\blacksquare \quad 327} \rightarrow 459$

d. $\underline{\blacksquare \quad 459} \rightarrow 327$

Part 2

- Here's a new type of problem:

Jerry weighed 72 pounds. Terry weighed 94 pounds. How much heavier is Terry than Jerry?

- The underlined sentence tells us that Terry is the big number. Jerry and box are the small numbers.

- Here's the number family: $\underline{\square \quad J} \rightarrow T$

- The first sentence tells us that Jerry is 72 and Terry is 94.

- We cross out Jerry and Terry and write their numbers.

- Here's the whole number family: $\underline{\square \quad \overset{72}{\cancel{J}}} \rightarrow \overset{94}{\cancel{T}}$

- Here's the subtraction problem and the answer:

- The unit name in the answer is pounds.

$$\begin{array}{r} 94 \\ -72 \\ \hline 22 \text{ pounds} \end{array}$$

Make a number family for each problem. Then write the addition or subtraction problem for each family and figure out the answer. Write the unit name in the answer.

a. Ginger weighed 85 pounds. Doris weighed 79 pounds. <u>How much heavier was Ginger than Doris?</u>

b. Sam ran 18 miles. Hank ran 26 miles. <u>How much farther did Hank run than Sam?</u>

c. Tina is 18 years old. Don is 37 years old. <u>How much younger is Tina than Don?</u>

d. Andrew is 77 inches tall. Fran is 60 inches tall. <u>How much shorter is Fran than Andrew?</u>

Part 3 **Paired Practice**

a. $8 - 6 = \blacksquare$ b. $10 - 6 = \blacksquare$ c. $7 - 5 = \blacksquare$ d. $13 - 9 = \blacksquare$

e. $4 - 2 = \blacksquare$ f. $4 - 3 = \blacksquare$ g. $12 - 8 = \blacksquare$ h. $12 - 10 = \blacksquare$

i. $12 - 12 = \blacksquare$ j. $11 - 7 = \blacksquare$ k. $5 - 4 = \blacksquare$ l. $7 - 4 = \blacksquare$

Independent Work

Part 4 **For each problem write a number family and figure out the answer. If there is a unit name, write it in the answer.**

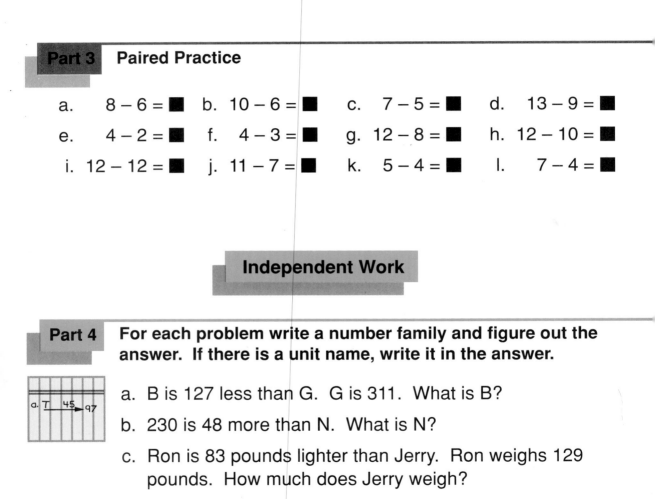

a. B is 127 less than G. G is 311. What is B?

b. 230 is 48 more than N. What is N?

c. Ron is 83 pounds lighter than Jerry. Ron weighs 129 pounds. How much does Jerry weigh?

d. Sara has 217 more dollars than Dave. Dave has 588 dollars. How many dollars does Sara have?

e. L is 450 less than R. L is 69. What is R?

f. G is 39 more than J. G is 148. What is J?

For each item draw a diagram. Then find the area.

a. A large rectangular field is 4 miles wide and 9 miles long. What is the area?

b. A piece of cardboard is the shape of a rectangle. The cardboard is 10 inches high and 8 inches wide. What is the area?

Part 6 **Write what X equals and what Y equals for each letter.**

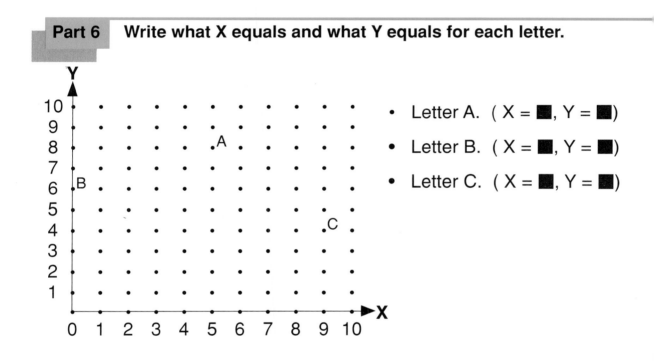

- Letter A. (X = ■, Y = ■)

- Letter B. (X = ■, Y = ■)

- Letter C. (X = ■, Y = ■)

Part 7 **Copy the table. Write the missing numbers.**

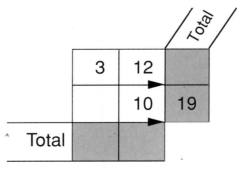

Lesson 52

Part 1

Make number families for each problem.
Then write the addition or subtraction problem for each
family and figure out the answer.
Write the unit name in the answer.

a. Benny weighs 110 pounds. Linda weighs 137 pounds.
How much lighter is Benny than Linda?

b. Rita has 235 nails. George has 45 nails. How many
more nails does Rita have than George has?

c. Donna is 11 years old. Benny is 46 years old. How
much younger is Donna than Benny?

d. The elm tree was 48 feet tall. The maple tree was 157
feet tall. How much shorter was the elm tree than the
maple tree?

Part 2

For each item, write the number family. Then write the
word **add** or **subtract** to tell what you would do to work
the problem.

a. 56 is 21 more than J.

b. Martha is 26 pounds heavier than Jan. Martha weighs
95 pounds.

c. 56 is 21 less than M.

d. The pole was 28 feet taller than the tree. The tree was
91 feet tall.

e. J is 18 more than 90.

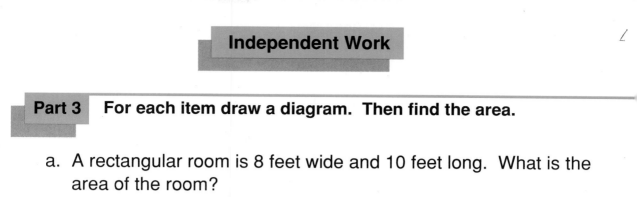

Independent Work

Part 3 For each item draw a diagram. Then find the area.

a. A rectangular room is 8 feet wide and 10 feet long. What is the area of the room?

b. A strip of paper is shaped like a rectangle. The paper is 9 inches long and 2 inches wide. What is the area of the paper?

Part 4 For each problem write a number family and figure out the answer.

a. F is 99 less than 274. What number is F?

b. R is 377 more than W. W is 248. What is R?

c. M is 117 less than P. P is 521. What is M?

Part 5 For each family write the multiplication fact that starts with the first small number.

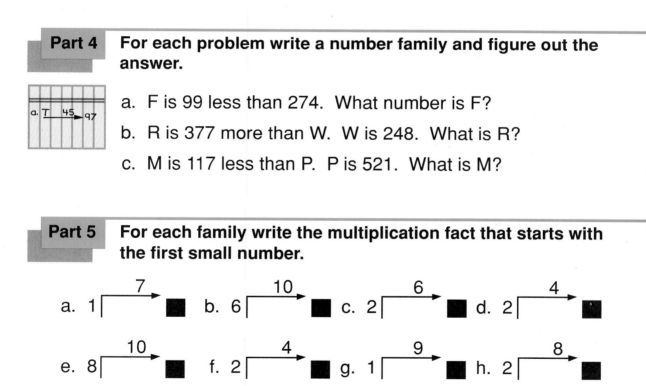

a. 1 ⟶ 7 ■   b. 6 ⟶ 10 ■   c. 2 ⟶ 6 ■   d. 2 ⟶ 4 ■

e. 8 ⟶ 10 ■   f. 2 ⟶ 4 ■   g. 1 ⟶ 9 ■   h. 2 ⟶ 8 ■

Part 6 Write the addition problem and the answer for each item.

a. You have 4 tens.   b. You have 3 fours.

Do the independent work for lesson 52 in your workbook.

Lesson 53

Part 1

- Hundreds numerals have **two** zeros.
- Tens numerals have **one** zero.
- The numeral for 51 **hundreds** is 51 and **two zeros.**
  - Here it is: **5100.**
- The numeral for 51 **tens** is 51 and **one zero.**
  - Here it is: **510.**

Turn your paper sideways. For each item, write the numeral.

a. 17 hundreds   b. 14 tens   c. 41 tens   d. 41 hundreds   e. 14 hundreds

Part 2

- In some of these problems, the **first sentence** tells how to make the number family.

- In some of these problems, the **last sentence** tells how to make the number family.

Make the number family for each problem.
Write the addition or subtraction problem for each family and figure out the answer.
Write the unit name in the answer.

a. Jane ran 11 miles farther than Ginger ran. Ginger ran 9 miles. How many miles did Jane run?

b. Jane rode a bike for 52 miles. Ginger rode for 36 miles. How much farther did Jane ride than Ginger?

c. Bill is 11 inches taller than Fran. Bill is 73 inches tall. How tall is Fran?

d. Al weighs 135 pounds. Janice weighs 110 pounds. How much heavier is Al than Janice?

Part 3

For each item, make a number family with two numbers and a letter. Then write **add** or **subtract** to tell what you would do to work the problem.

a. B is 18 less than 90.

b. The box held 45 pounds less than the tub. The box held 70 pounds.

c. 59 is 17 more than B.

d. J is 90 more than B. B is 175.

e. R is 12 less than 96.

Part 4 For each family write the multiplication fact that starts with the **second** small number.

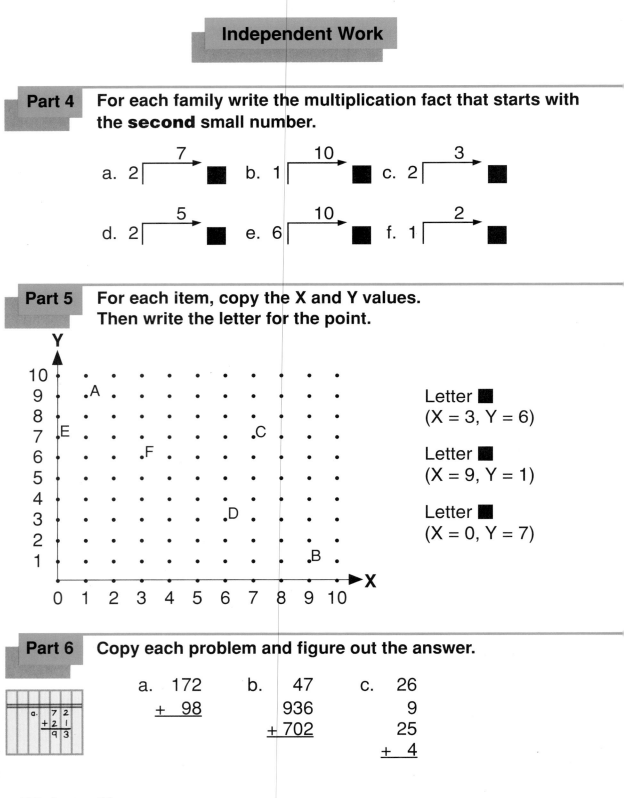

a. 2⌐7→ ▪   b. 1⌐10→ ▪   c. 2⌐3→ ▪

d. 2⌐5→ ▪   e. 6⌐10→ ▪   f. 1⌐2→ ▪

Part 5 For each item, copy the X and Y values.
Then write the letter for the point.

Letter ▪
(X = 3, Y = 6)

Letter ▪
(X = 9, Y = 1)

Letter ▪
(X = 0, Y = 7)

Part 6 Copy each problem and figure out the answer.

a.	172	b.	47	c.	26
	+ 98		936		9
			+ 702		25
					+ 4

Part 7 Write the numbers that could be F for each item.

a. F ⟩ 1 | 1000 10 7 0 | b. F ⟨ 100 | 99 101 0 1000 |

Part 8 Copy each problem and write the answer.

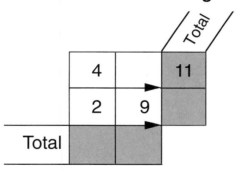

a. 927   b. 280   c. 1609   d. 765
  − 649   − 174   − 390   − 196

Part 9 Copy the table. Write the missing numbers.

		Total
4		11
2	9	
Total		

Lesson 54

- For some problems, you don't know whether they are addition problems or subtraction problems until you put the values in a number family.

- Here are three values: First value: 26
                          Second value: 12
                          Third value: ☐

- We'll put the first value in the number family first, the second value in second, and the third value in third.

- For some problems, we'll put the values in forward along the arrow.

- For other problems, we'll put the values in backward along the arrow.

- Here are the three values written **forward** along the number family arrow. The first value is the first small number.

- Here are the same values written **backward** along the number family arrow. Now the first value is the big number.

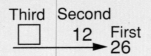

- Remember, if you put the values in backward along the arrow, the first value is the big number.

a.	b.	c.	d.	e.
57	57	□	80	38
19	19	19	19	18
□	□	80	□	□

Part 2   **Write answers to all the problems.**

a. $62 + 5 = $ ■   b. $34 + 3 = $ ■   c. $26 + 2 = $ ■

d. $42 + 7 = $ ■   e. $93 + 5 = $ ■

Part 3   **Make a complete number family for each problem.**
Then write the addition problem or subtraction problem for
each family and figure out the answer.
Remember the unit names.

a. An elm tree was 15 feet shorter than a pine tree. The
   pine tree was 84 feet tall. How tall was the elm tree?

b. Alice ran 17 miles farther than Rita. Rita ran 16 miles.
   How far did Alice run?

c. Don had 78 more eggs than George had. Don had 268
   eggs. How many eggs did George have?

d. Fran ran 17 miles. Rita ran 6 miles. How many more
   miles did Fran run than Rita ran?

e. An elm tree was 15 feet tall. A pine tree was 84 feet tall.
   How much shorter was the elm tree than the pine tree?

Part 4 Paired Practice

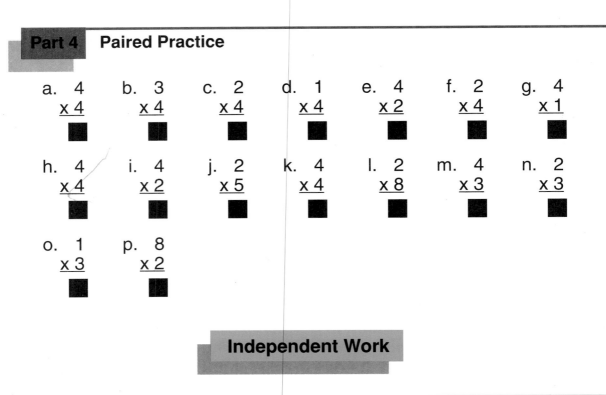

a. 4
  x 4

b. 3
  x 4

c. 2
  x 4

d. 1
  x 4

e. 4
  x 2

f. 2
  x 4

g. 4
  x 1

h. 4
  x 4

i. 4
  x 2

j. 2
  x 5

k. 4
  x 4

l. 2
  x 8

m. 4
  x 3

n. 2
  x 3

o. 1
  x 3

p. 8
  x 2

Independent Work

Part 5   For each item, make a number family with two numbers and a letter.
Then write **add** or **subtract** to tell what you would do to work the problem.

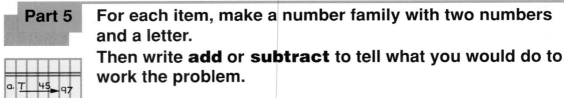

a. The pole is 41 inches shorter than the tree. The tree is 262 inches tall.

b. 87 is 46 less than K.

c. The alligator was 12 pounds heavier than the turtle. The alligator weighed 39 pounds.

d. J is 94 more than K. J is 118.

e. 200 is 90 more than R.

f. J is 300 less than K. K is 301.

Part 6 Write what X equals and what Y equals for each letter.

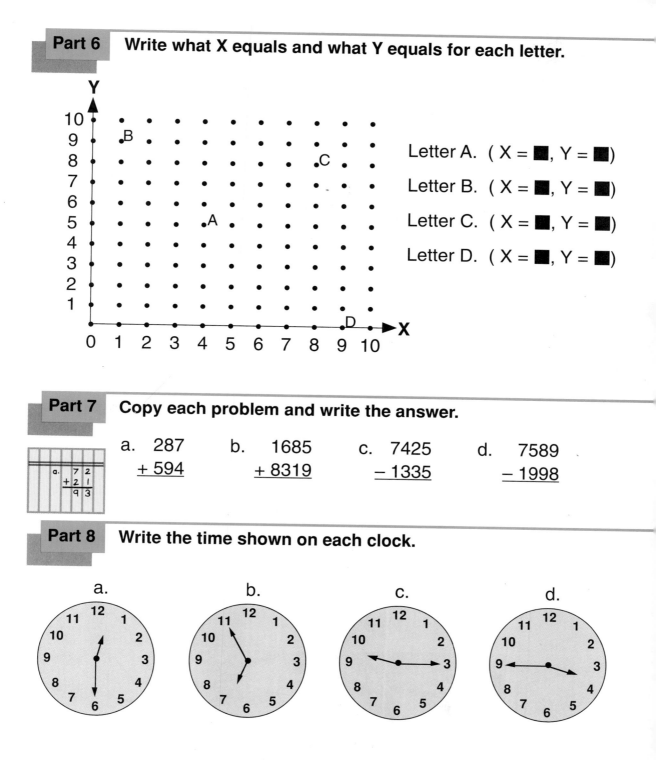

Letter A. (X = ■, Y = ■)

Letter B. (X = ■, Y = ■)

Letter C. (X = ■, Y = ■)

Letter D. (X = ■, Y = ■)

Part 7 Copy each problem and write the answer.

a. 287
   + 594

b. 1685
   + 8319

c. 7425
   − 1335

d. 7589
   − 1998

Part 8 Write the time shown on each clock.

a.

b.

c.

d.

Write the answer to each question.

The table shows the number of red birds and yellow birds that were seen on Monday and Tuesday.

	Monday	Tuesday	Total for both days
Red birds	7	4	11
Yellow birds	1	14	15
Total birds	8	18	

a. How many yellow birds were seen on Tuesday?

b. Were more total birds seen on Monday or on Tuesday?

c. How many red birds were seen on Monday?

d. How many yellow birds were seen on both days?

Do the independent work for lesson 54 in your workbook.

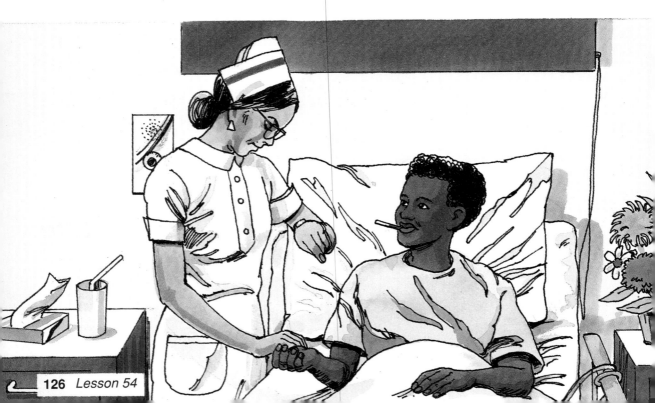

Part 1 Make the number family for each problem.
Write the addition or subtraction problem and figure out
the answer.
Write the unit name in the answer.

a. Rita had 56 feet of string. Al had 19 feet of string. How much less string did Al have than Rita had?

b. Fran walked 12 more miles than Al walked. Al walked 14 miles. How many miles did Fran walk?

c. Donna weighed 22 pounds less than Eric weighed. Eric weighed 150 pounds. How much did Donna weigh?

d. A car was 24 feet long. A truck was 59 feet long. How much shorter was the car than the truck?

Part 2

a. Put these values in **backward** along the number family arrow.

First value: ☐
Second value: 14
Third value: 13

b. Put these values in **forward** along the number family arrow.

First value: ☐
Second value: 14
Third value: 19

c. Put these values in **forward** along the number family arrow.

First value: 13
Second value: ☐
Third value: 18

d. Put these values in **backward** along the number family arrow.

First value: 33
Second value: 28
Third value: ☐

e. Put these values in **forward**
   along the number family arrow.

   First value: 33
   Second value: 28
   Third value: ☐

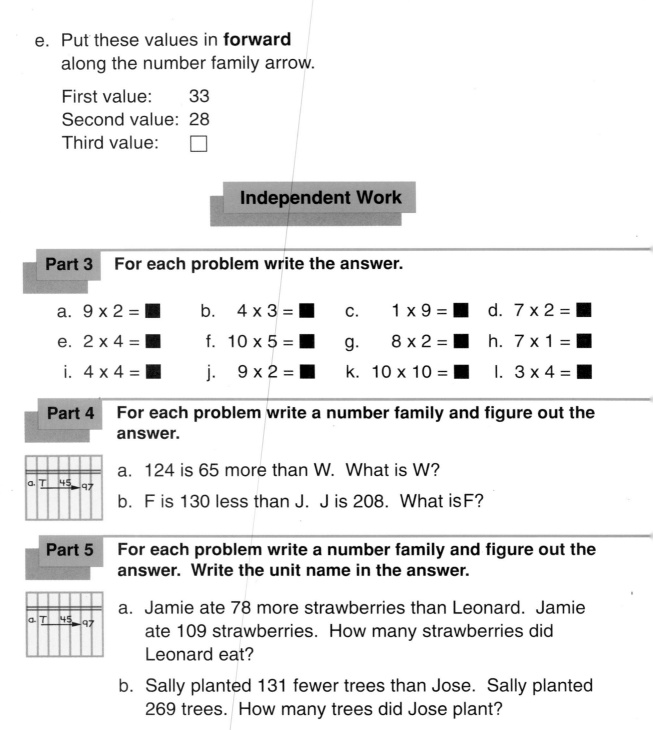

Independent Work

Part 3   **For each problem write the answer.**

a. 9 x 2 = ■   b. 4 x 3 = ■   c.   1 x 9 = ■   d. 7 x 2 = ■

e. 2 x 4 = ■   f. 10 x 5 = ■   g.   8 x 2 = ■   h. 7 x 1 = ■

i. 4 x 4 = ■   j.   9 x 2 = ■   k. 10 x 10 = ■   l. 3 x 4 = ■

Part 4   **For each problem write a number family and figure out the answer.**

a. 124 is 65 more than W. What is W?

b. F is 130 less than J. J is 208. What is F?

Part 5   **For each problem write a number family and figure out the answer. Write the unit name in the answer.**

a. Jamie ate 78 more strawberries than Leonard. Jamie ate 109 strawberries. How many strawberries did Leonard eat?

b. Sally planted 131 fewer trees than Jose. Sally planted 269 trees. How many trees did Jose plant?

c. Kurt wrote 304 fewer letters than Andrea. Andrea wrote 824 letters. How many letters did Kurt write?

Part 6 Copy the table. Write the missing numbers.

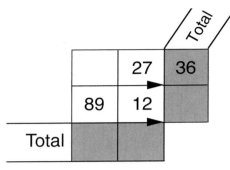

	27	**36** Total
89	12	
Total		

Do the independent work for lesson 55 in your workbook.

Lesson 56

Part 1

a. Put these values in **forward** along the number family arrow.

First value: ☐
Second value: 12
Third value 56

b. Put these values in **backward** along the number family arrow.

First value: ☐
Second value: 12
Third value 56

c. Put these values in **forward** along the number family arrow.

First value: 15
Second value: 13
Third value ☐

d. Put these values in **backward** along the number family arrow.

First value: 15
Second value: 13
Third value ☐

Part 2 Make a complete number family for each problem.
Write the addition or subtraction problem and figure out the answer.
Write the unit name in the answer.

a. A truck weighed 1000 pounds more than a car. The truck weighed 5100 pounds. How much did the car weigh?

b. A bus was 14 feet wide. A house was 55 feet wide. How much wider was the house than the bus?

c. A worm was 23 inches shorter than a snake. The worm was 9 inches long. How long was the snake?

d. A turtle was 96 years old. A dog was 14 years old. How much younger was the dog than the turtle?

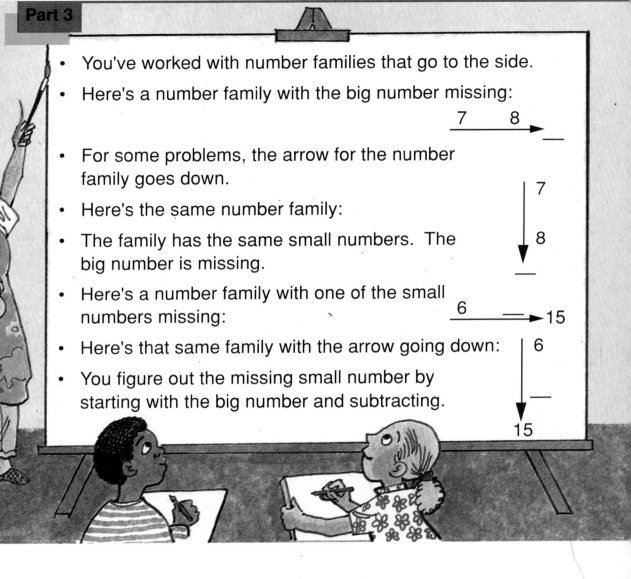

- You've worked with number families that go to the side.
- Here's a number family with the big number missing:

7   8

- For some problems, the arrow for the number family goes down.
- Here's the same number family:
- The family has the same small numbers. The big number is missing.

7

8

- Here's a number family with one of the small numbers missing:

6 =→ 15

- Here's that same family with the arrow going down:

6

15

- You figure out the missing small number by starting with the big number and subtracting.

For each problem, write the addition or subtraction problem and the answer.

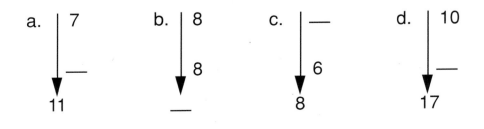

a. 7

   —

   11

b. 8

   8

   —

c. —

   6

   8

d. 10

   —

   17

Part 4   For each problem write a number family and figure out the answer.

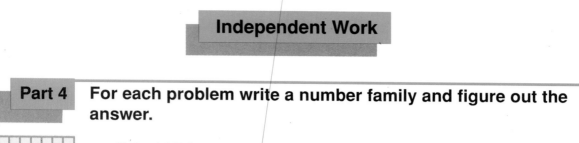

a. B is 127 less than G. G is 311. What is B?

b. 280 is 68 more than N. What is N?

c. Ron is 83 pounds lighter than Gail. Ron weighs 129 pounds. What does Gail weigh?

d. Sara has 217 more socks than Dave. Dave has 588 socks. How many socks does Sara have?

Part 5   For each problem write the answer.

a.   2 x 10 = ■   b. 4 x 4 = ■   c. 9 x 1 = ■   d. 8 x 2 = ■

e.   3 x 4 = ■   f. 4 x 1 = ■   g. 4 x 4 = ■   h. 2 x 7 = ■

i. 10 x 10 = ■   j. 2 x 4 = ■   k. 9 x 3 = ■   l. 6 x 2 = ■

Do the independent work for lesson 56 in your workbook.

Lesson 57

Part 1

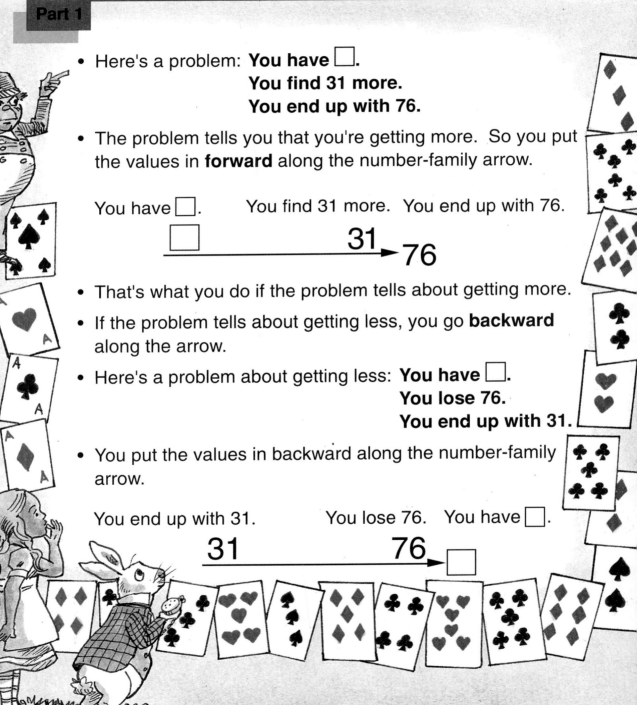

- Here's a problem: **You have ☐.**
  You find 31 more.
  You end up with 76.

- The problem tells you that you're getting more. So you put the values in **forward** along the number-family arrow.

You have ☐.   You find 31 more. You end up with 76.

$$\boxed{} \xrightarrow{\hspace{2cm} 31 \hspace{0.5cm}} 76$$

- That's what you do if the problem tells about getting more.

- If the problem tells about getting less, you go **backward** along the arrow.

- Here's a problem about getting less: **You have ☐.**
  You lose 76.
  You end up with 31.

- You put the values in backward along the number-family arrow.

You end up with 31.   You lose 76.   You have ☐.

$$31 \xrightarrow{\hspace{4cm} 76 \hspace{0.3cm}} \boxed{}$$

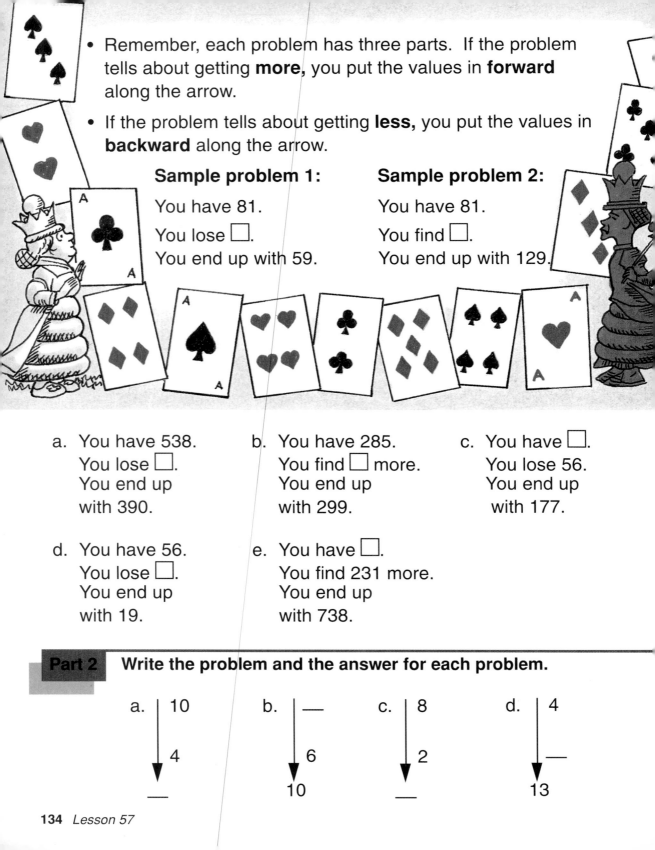

- Remember, each problem has three parts. If the problem tells about getting **more,** you put the values in **forward** along the arrow.

- If the problem tells about getting **less,** you put the values in **backward** along the arrow.

Sample problem 1:

You have 81.

You lose ☐.

You end up with 59.

Sample problem 2:

You have 81.

You find ☐.

You end up with 129.

a. You have 538.
   You lose ☐.
   You end up
   with 390.

b. You have 285.
   You find ☐ more.
   You end up
   with 299.

c. You have ☐.
   You lose 56.
   You end up
   with 177.

d. You have 56.
   You lose ☐.
   You end up
   with 19.

e. You have ☐.
   You find 231 more.
   You end up
   with 738.

Part 2   Write the problem and the answer for each problem.

a.   10
     ↓
     4
     ___

b.   ___
     ↓
     6
     10

c.   8
     ↓
     2
     ___

d.   4
     ↓
     ___
     13

- Each fraction has two numbers–a top number and a bottom number. The **bottom** number tells the number of parts each inch is divided into.

- The **bottom** number is the same for the all the fractions on a number line.

- The **top** number tells the numbers of parts from the beginning of the number line.

- The **top** number is different for each fraction on the number line.

- Here's a picture of a number line with empty fractions at each inch:

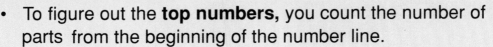

- To write out the fractions you figure out the **bottom numbers first.**

- Each inch is divided into 3 parts. So the bottom number of each fraction is 3.

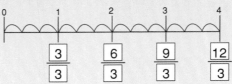

- To figure out the **top numbers,** you count the number of parts from the beginning of the number line.

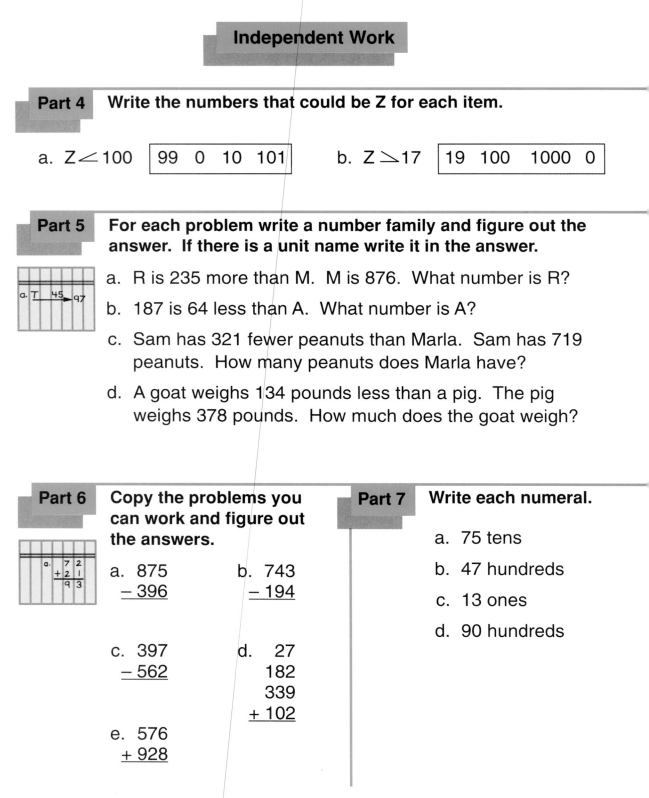

Independent Work

Part 4 Write the numbers that could be Z for each item.

a. Z < 100 | 99 0 10 101 | b. Z ≥ 17 | 19 100   1000 0 |

Part 5 For each problem write a number family and figure out the answer. If there is a unit name write it in the answer.

a. R is 235 more than M. M is 876. What number is R?

b. 187 is 64 less than A. What number is A?

c. Sam has 321 fewer peanuts than Marla. Sam has 719 peanuts. How many peanuts does Marla have?

d. A goat weighs 134 pounds less than a pig. The pig weighs 378 pounds. How much does the goat weigh?

Part 6 Copy the problems you can work and figure out the answers.

a. 875
  − 396

b. 743
  − 194

c. 397
  − 562

d.   27
   182
   339
  + 102

e. 576
  + 928

Part 7 Write each numeral.

a. 75 tens

b. 47 hundreds

c. 13 ones

d. 90 hundreds

Part 8

Write the addition problem or the subtraction problem and figure out the answer.

a. A woman had $4.17. She spent $2.09. How much money did she end up with?

b. A woman had $4.17. Then she spent $4.17. How much money did she end up with?

c. A man had $14.00. Then he spent $7.00. How much money did he end up with?

Part 9

Write the time shown on each clock.

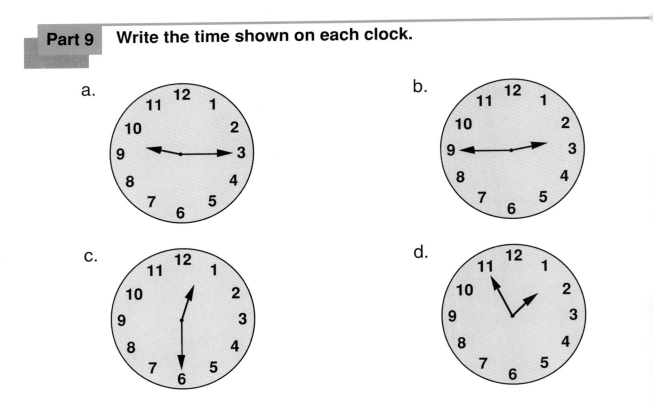

a.

b.

c.

d.

Part 1 Make a number family for each problem.
Write the addition or subtraction problem and figure out
the answer for each family.

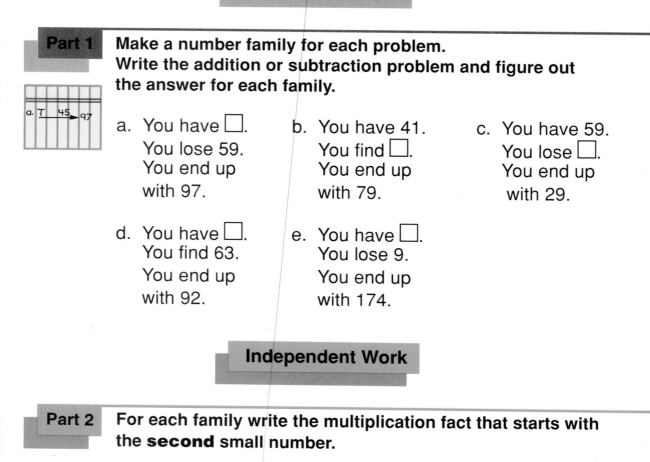

a. You have ☐.
   You lose 59.
   You end up
   with 97.

b. You have 41.
   You find ☐.
   You end up
   with 79.

c. You have 59.
   You lose ☐.
   You end up
   with 29.

d. You have ☐.
   You find 63.
   You end up
   with 92.

e. You have ☐.
   You lose 9.
   You end up
   with 174.

Independent Work

Part 2 For each family write the multiplication fact that starts with
the **second** small number.

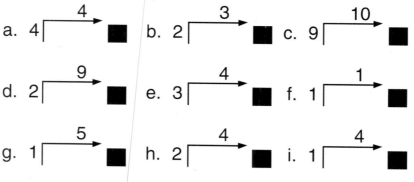

For each problem write a number family and figure out the answer. Remember the unit name.

a. An oak tree is 51 feet shorter than a pine tree. The oak tree is 39 feet tall. How tall is the pine tree?

b. A boat was 32 feet long. A truck was 61 feet long. How much longer was the truck than the boat?

c. A man weighed 234 pounds. A goat weighed 195 pounds. How much heavier was the man than the goat?

d. A car went 293 miles farther than a truck. The car went 589 miles. How far did the truck go?

Part 4   Copy the problems you can work and figure out the answers.

a. 491
+ 530

b. 530
− 491

c. 491
− 530

d. 566
− 729

e. 396
− 281

f. 427
+ 400

Part 5   Write each numeral.

a. 14 hundreds

b. 16 ones

c. 13 tens

d. 40 tens

e. 30 hundreds

Lesson 59

- When you multiply by tens numbers, your answer is a tens number. So it has a zero in the ones column.

- Here's 5 times 3. The answer is 15.

$$\begin{array}{r} 5 \\ \times\ 3 \\ \hline 15 \end{array}$$

- Here's 50 times 3. That's 5 **tens** times 3. The answer is 15 **tens.** That's 150.

$$\begin{array}{r} 50 \\ \times\ \ 3 \\ \hline 150 \end{array}$$

- Here's 2 times 4. That's 8.

$$\begin{array}{r} 2 \\ \times\ 4 \\ \hline 8 \end{array}$$

- Here's 20 times 4. That's 2 **tens** times 4. The answer is 8 **tens.** That's 80.

$$\begin{array}{r} 20 \\ \times\ 4 \\ \hline 80 \end{array}$$

- Here's 4 times 5. That's 20.

$$\begin{array}{r} 4 \\ \times\ 5 \\ \hline 20 \end{array}$$

- Here's 40 time 5. That's 4 **tens** times 5. The answer is 20 **tens.** That's 200.

$$\begin{array}{r} 40 \\ \times\ \ 5 \\ \hline 200 \end{array}$$

a.
$$\begin{array}{r}4 \\ \times\ 3 \\ \hline \end{array}\qquad \begin{array}{r}40 \\ \times\ \ 3 \\ \hline \end{array}$$

b.
$$\begin{array}{r}2 \\ \times\ 3 \\ \hline \end{array}\qquad \begin{array}{r}20 \\ \times\ \ 3 \\ \hline \end{array}$$

c.
$$\begin{array}{r}5 \\ \times\ 3 \\ \hline \end{array}\qquad \begin{array}{r}50 \\ \times\ \ 3 \\ \hline \end{array}$$

d.
$$\begin{array}{r}9 \\ \times\ 3 \\ \hline \end{array}\qquad \begin{array}{r}90 \\ \times\ \ 3 \\ \hline \end{array}$$

e.
$$\begin{array}{r}4 \\ \times\ 5 \\ \hline \end{array}\qquad \begin{array}{r}40 \\ \times\ \ 5 \\ \hline \end{array}$$

Part 2

a. You have ☐.
You find 45.
You end up
with 184.

b. You have ☐.
You lose 184.
You end up
with 426.

c. You have 198.
You lose ☐.
You end up
with 79.

d. You have 207.
You find ☐.
You end up
with 229.

e. You have ☐.
You lose 320.
You end up
with 52.

Part 3 Paired Practice

a. $10 - 6 = \blacksquare$ b. $8 - 6 = \blacksquare$ c. $12 - 6 = \blacksquare$ d. $9 - 6 = \blacksquare$

e. $11 - 6 = \blacksquare$ f. $7 - 6 = \blacksquare$ g. $11 - 6 = \blacksquare$ h. $10 - 6 = \blacksquare$

i. $9 - 6 = \blacksquare$ j. $12 - 6 = \blacksquare$ k. $8 - 6 = \blacksquare$ l. $9 - 6 = \blacksquare$

m. $10 - 6 = \blacksquare$ n. $11 - 6 = \blacksquare$ o. $12 - 6 = \blacksquare$

Part 4 Write what X equals and what Y equals for each letter.

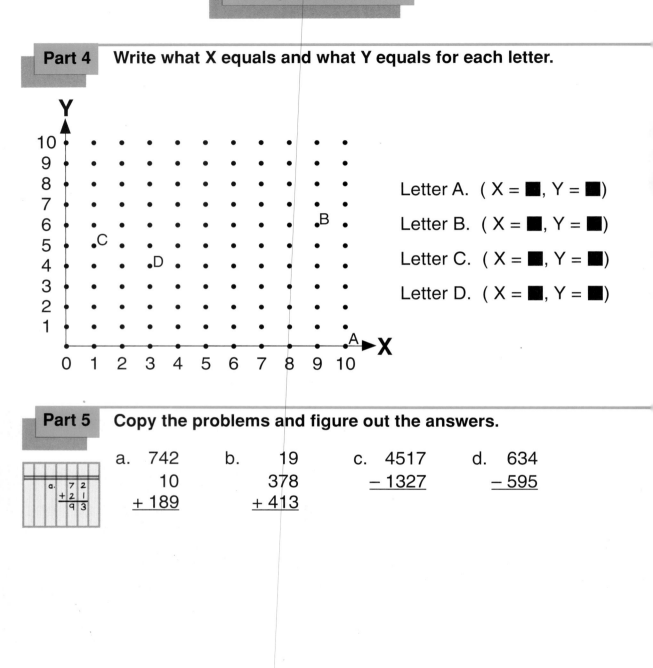

Letter A. (X = ■, Y = ■)

Letter B. (X = ■, Y = ■)

Letter C. (X = ■, Y = ■)

Letter D. (X = ■, Y = ■)

Part 5 Copy the problems and figure out the answers.

a.   742   b.   19   c.   4517   d.   634
      10   378   − 1327   − 595
   + 189   + 413

Lesson 60

Part 1

- You've worked table problems by finding the missing number in rows. Each row works just like a number family. The first two numbers are small numbers.

- The number for the total is the big number.

- The columns work the same way, except the number family goes down.

- The top two numbers are the small numbers.

- The bottom number is the big number.

		Total
3	4	7
8	2	10
Total 11	6	17

		Total
3	4	7
8	2	10
Total 11	6	17

- Here's the rule about rows and columns: If there are two numbers in a **row,** you can figure out the missing number. If there is only one number in a row, you can't figure out the missing number.

- If there are two numbers in a **column,** you can figure out the missing number. If there's only one number in a column, you can't figure out the missing number.

**Make a number family for each problem. Remember, just
make a box for some.
Write the addition or subtraction problem and figure out
the answer.**

a. You have some.
   You find 36.
   You end up
   with 56.

b. You have some.
   You lose 58.
   You end up
   with 79.

c. You have 41.
   You lose some.
   You end up
   with 29.

d. You have some.
   You find 57.
   You end up
   with 86.

e. You have some.
   You lose 49.
   You end up
   with 56.

**For each problem, make a number family.
Write the column problem and figure out the answer.
If there is a unit name write it in the answer.**

a. R is 167 more than M. R is 229. What is M?

b. Rob was 63 inches tall. Debbie was 71 inches tall. How
   much shorter was Rob than Debbie?

c. Joe weighs 39 pounds less than Mike. Joe weighs 168
   pounds. How much does Mike weigh?

d. A pig weighed 136 pounds more than a fox. The fox
   weighed 59 pounds. How much did the pig weigh?

Part 4 For each item, write the numbers in the box that could be the letter.

a. 10∠Z | 0   100   10   11 |

c. 11 ⊃ T | 10   0   11   3 |

b. T∠72 | 100   69   0   72 |

Part 5 Write each numeral.

a. 16 ones

b. 40 tens

c. 50 hundreds

d. 8 hundreds

e. 17 hundreds

f. 4 tens

Part 6 Write the time shown on each clock.

a.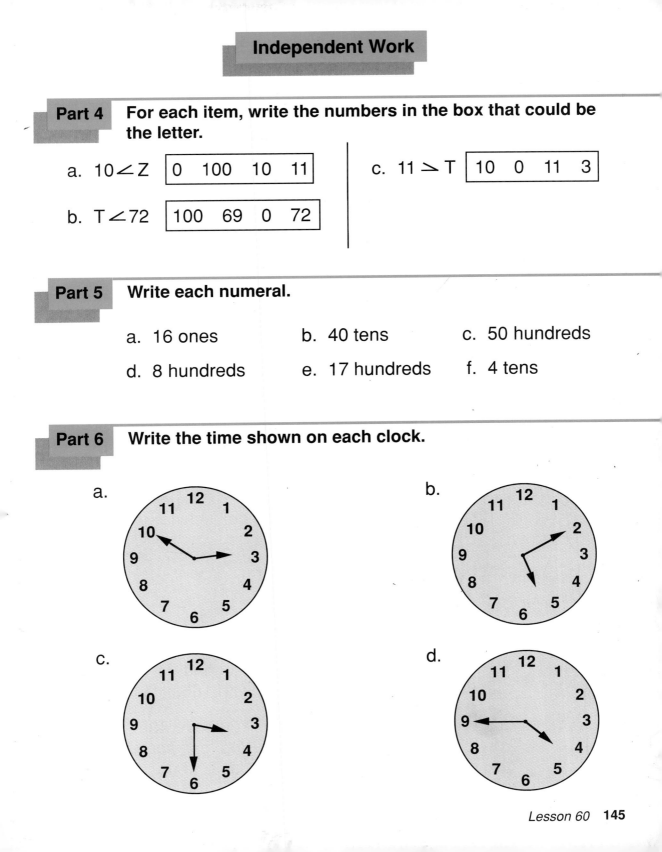

b.

c.

d.

Lesson 61

Part 1

a. Rita had 56 feet of string. Al had 19 feet of string. How much less string did Al have than Rita had?

b. Sid walked 14 miles. Fran walked 12 more miles than Sid walked. How many miles did Fran walk?

c. Sid weighed 150 pounds. Donna weighed 22 pounds less than Sid weighed. How much did Donna weigh?

d. A car was 24 feet long. A truck was 59 feet long. How much shorter was the car than the truck?

Part 2

- These problems have the word **some.**

- You write a box for **some** because some doesn't tell you the number.

- Remember, if you get less, the values go backward along the arrow.

- If you get more, the values go forward.

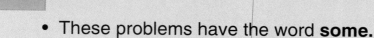

a. You have some.
   You lose 18.
   You end up
   with 68.

b. You have 7.
   You find some.
   You end up
   with 166.

c. You have some.
   You lose 169.
   You end up
   with 232.

d. You have 488.
   You lose some.
   You end up
   with 406.

Paired Practice

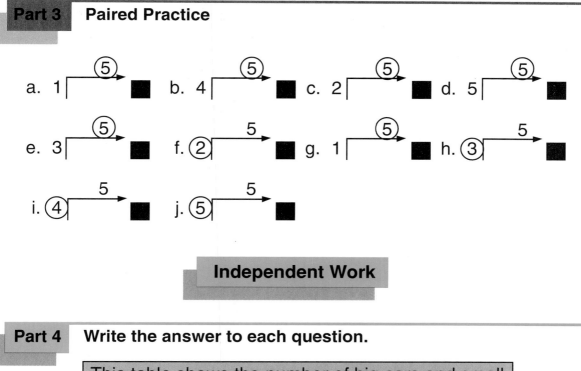

a. 1 $\xrightarrow{\textcircled{5}}$ ■ b. 4 $\xrightarrow{\textcircled{5}}$ ■ c. 2 $\xrightarrow{\textcircled{5}}$ ■ d. 5 $\xrightarrow{\textcircled{5}}$ ■

e. 3 $\xrightarrow{\textcircled{5}}$ ■ f. ② $\xrightarrow{5}$ ■ g. 1 $\xrightarrow{\textcircled{5}}$ ■ h. ③ $\xrightarrow{5}$ ■

i. ④ $\xrightarrow{5}$ ■ j. ⑤ $\xrightarrow{5}$ ■

Independent Work

Part 4 **Write the answer to each question.**

> This table shows the number of big cars and small cars that parked in lot A and B.

	Lot A	Lot B	Total for both lots
Big cars	13	12	25
Small cars	16	14	30
Total cars	29	26	

a. How many big cars were there in both lots?

b. How many small cars parked in lot A?

c. How many big cars parked in lot B?

d. Were there more small cars or big cars parked in lot B?

Part 5 **Copy the problem and figure out the answer.**

```
 7 2
+2 1
 9 3
```

a. 507
  − 426

b.   76
    138
  + 18

c. 111
     19
  + 480

d. 741
  − 295

Write what X equals and what Y equals for each letter.

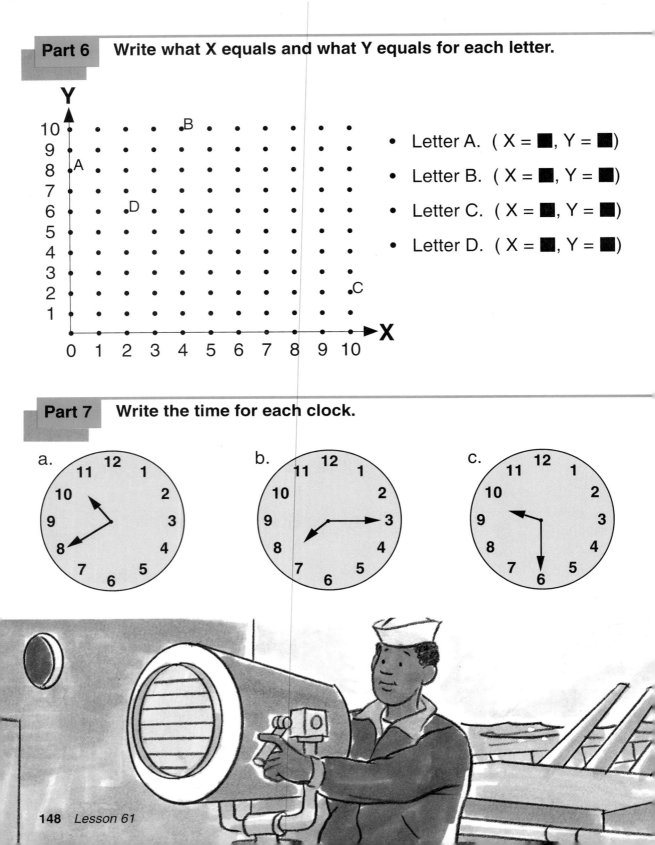

- Letter A. (X = ■, Y = ■)

- Letter B. (X = ■, Y = ■)

- Letter C. (X = ■, Y = ■)

- Letter D. (X = ■, Y = ■)

Part 7 **Write the time for each clock.**

a.

b.

c.

Lesson 62

Part 1

- If you get more, the values go forward along the arrow.
- If you get less, the values go backward along the arrow.
- You write a box for **some.**

Make a number family for each problem. Write the addition or subtraction problem and figure out the answer. Remember the unit name.

a. Tim had 36 dimes. Then he earned some more dimes. He ended up with 89 dimes. How many dimes did he earn?

b. A dog started out with some fleas. Then the dog got rid of 48 fleas. The dog ended up with 99 fleas. How many fleas did the dog start out with?

c. A tree was 16 feet tall. Then the tree grew some more. The tree ended up being 99 feet tall. How many feet did the tree grow?

d. Sam had 154 buttons. Then he gave away some buttons. He ended up with 64 buttons. How many buttons did he give away?

Part 2   Write all the numbers for counting by 9 to 90.

9 ▪ ▪ ▪ ▪ ▪ ▪ ▪ ▪ 90

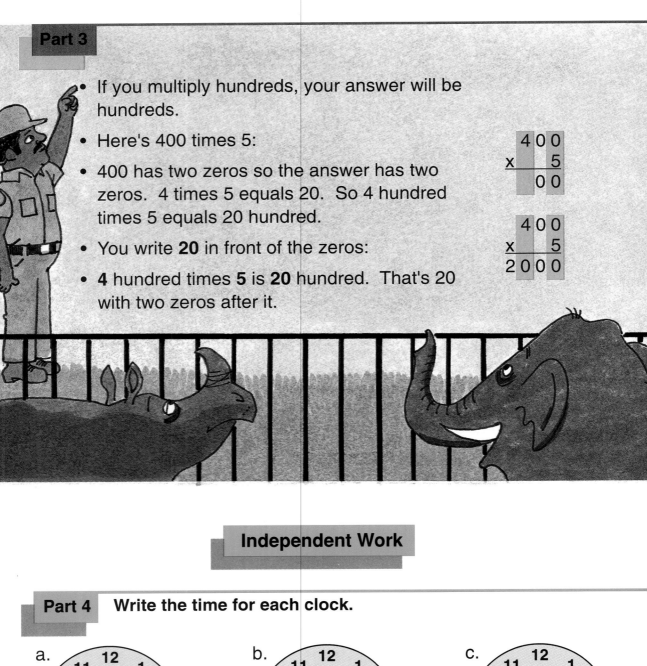

- If you multiply hundreds, your answer will be hundreds.

- Here's 400 times 5:

- 400 has two zeros so the answer has two zeros. 4 times 5 equals 20. So 4 hundred times 5 equals 20 hundred.

- You write **20** in front of the zeros:

- **4** hundred times **5** is **20** hundred. That's 20 with two zeros after it.

```
    4 0 0
x       5
      0 0
```

```
    4 0 0
x       5
  2 0 0 0
```

Independent Work

Part 4 Write the time for each clock.

a.

b.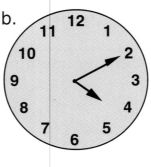

c.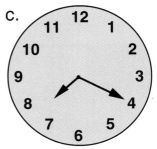

Part 5   **Answer the questions.**

This table shows the number of people who visited Rainbow Valley and Big Lake in the fall and winter.

	Fall	Winter	Total
Rainbow Valley	106	54	160
Big Lake	206	56	262
Total	312	110	422

a. How many total people visited Big Lake in the winter and the fall?

b. How many people visited Rainbow Valley in the fall?

c. How many total people visited both places in the fall?

d. How many total people visited Rainbow Valley?

Part 6   **Copy each problem and write the answer.**

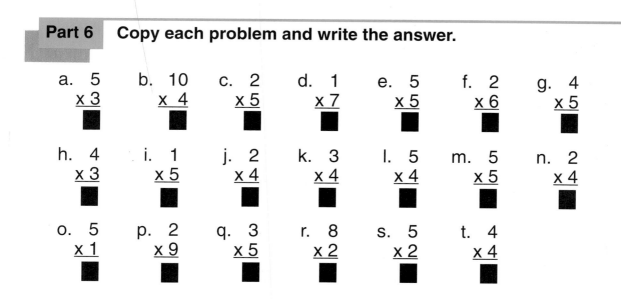

a.   5
   x 3
   ■

b.   10
   x 4
   ■

c.   2
   x 5
   ■

d.   1
   x 7
   ■

e.   5
   x 5
   ■

f.   2
   x 6
   ■

g.   4
   x 5
   ■

h.   4
   x 3
   ■

i.   1
   x 5
   ■

j.   2
   x 4
   ■

k.   3
   x 4
   ■

l.   5
   x 4
   ■

m.   5
   x 5
   ■

n.   2
   x 4
   ■

o.   5
   x 1
   ■

p.   2
   x 9
   ■

q.   3
   x 5
   ■

r.   8
   x 2
   ■

s.   5
   x 2
   ■

t.   4
   x 4
   ■

Lesson 63

Part 1 Make the number family for each problem.
Write the number problem and figure out the answer.
Remember the unit name.

a. Jenny started out with some cans. She got 36 more
cans. She ended up with 58 cans. How many cans did
she start with?

b. Sam had 430 bolts. Then he used some bolts. He
ended up with 330 bolts. How many bolts did Sam use?

c. Jane had some bolts. She got 130 more bolts. She
ended up with 520 bolts. How many bolts did Jane start
with?

d. Jose had 95 cans. Rita had 14 more cans than Jose
had.   How many cans did Rita have?

e. Henry had 150 more bolts than Jane had. Henry had
561 bolts. How many bolts did Jane have?

Independent Work

Part 2 Copy each problem and write the answer.

a. $9 - 3 = $ 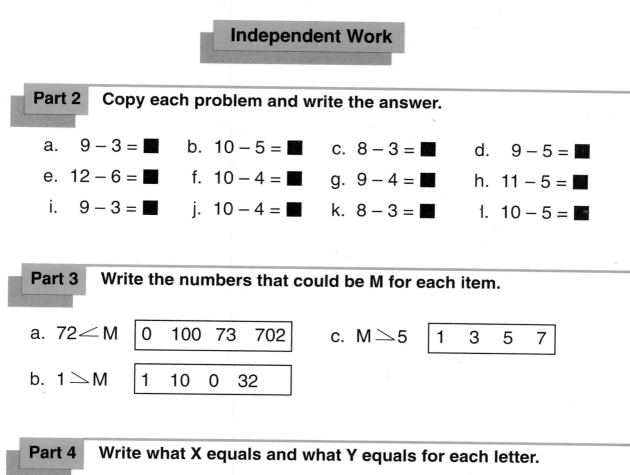 b. $10 - 5 = $ c. $8 - 3 = $ d. $9 - 5 = $

e. $12 - 6 = $ f. $10 - 4 = $ g. $9 - 4 = $ h. $11 - 5 = $

i. $9 - 3 = $ j. $10 - 4 = $ k. $8 - 3 = $ l. $10 - 5 = $

Part 3 Write the numbers that could be M for each item.

a. $72 < M$ | 0 100 73 702 | c. $M \geq 5$ | 1 3 5 7 |

b. $1 \geq M$ | 1 10 0 32 |

Part 4 Write what X equals and what Y equals for each letter.

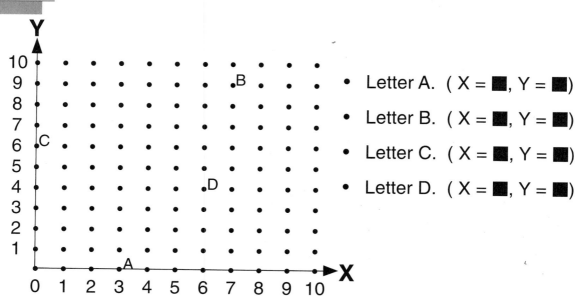

- Letter A. (X = ■, Y = ■)
- Letter B. (X = ■, Y = ■)
- Letter C. (X = ■, Y = ■)
- Letter D. (X = ■, Y = ■)

Lesson 63 **153**

Lesson 64

Part 1

Make the number family for each problem.
Write the number problem and figure out the answer.
Remember the unit name.

a. A cow ate 37 more carrots than a goat ate. The cow ate 148 carrots. How many carrots did the goat eat?

b. Rita's rocks weighed 137 pounds. Rita's rocks weighed 101 pounds less than Henry's rocks. How much did Henry's rocks weigh?

c. Henry had 146 rocks. He threw away some rocks. He ended up with 50 rocks. How many rocks did he throw away?

d. Rita had 51 rocks. Then she found 37 more rocks. How many rocks did she end up with?

Part 2

- In some multiplication problems, the top number ends in zeros.

- When you work these problems, you first copy the zeros in the answer. Then you figure out the first digits of the answer.

- If the top number ends in one zero, write one zero at the end of the answer.

$$\begin{array}{r} 5\,0 \\ \times\ \ 4 \\ \hline 2\,0\,0 \end{array}$$

Copy each problem and work it.

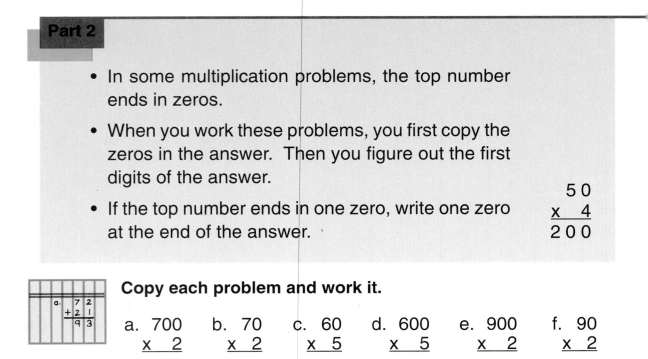

a. $\begin{array}{r} 700 \\ \times\ \ 2 \\ \hline \end{array}$
b. $\begin{array}{r} 70 \\ \times\ 2 \\ \hline \end{array}$
c. $\begin{array}{r} 60 \\ \times\ 5 \\ \hline \end{array}$
d. $\begin{array}{r} 600 \\ \times\ \ 5 \\ \hline \end{array}$
e. $\begin{array}{r} 900 \\ \times\ \ 2 \\ \hline \end{array}$
f. $\begin{array}{r} 90 \\ \times\ 2 \\ \hline \end{array}$

a. $12 - 6 = \blacksquare$ b. $10 - 4 = \blacksquare$ c. $10 - 6 = \blacksquare$ d. $8 - 6 = \blacksquare$

e. $11 - 6 = \blacksquare$ f. $9 - 6 = \blacksquare$ g. $11 - 5 = \blacksquare$ h. $9 - 3 = \blacksquare$

i. $8 - 2 = \blacksquare$ j. $12 - 6 = \blacksquare$ k. $7 - 6 = \blacksquare$ l. $9 - 4 = \blacksquare$

m. $9 - 6 = \blacksquare$ n. $10 - 6 = \blacksquare$ o. $9 - 3 = \blacksquare$

Independent Work

Part 4   **Answer the questions.**

This table shows the number of cows and horses in the barn and in the field.

	Cows	Horses	Total for both animals
Field	9	1	10
Barn	7	2	9
Total for both places	16	3	19

a. How many horses are in the field?

b. How many cows and horses are in the barn?

c. How many cows are in both places?

d. How many cows and horses are in the field?

Part 5 Copy each problem and write the answer.

a. 7
 x 1
 ■

b. 6
 x 5
 ■

c. 10
 x 4
 ■

d. 3
 x 4
 ■

e. 9
 x 2
 ■

f. 8
 x 5
 ■

g. 10
 x 1
 ■

h. 2
 x 3
 ■

i. 4
 x 7
 ■

j. 5
 x 4
 ■

Part 6 Copy each problem and work it.

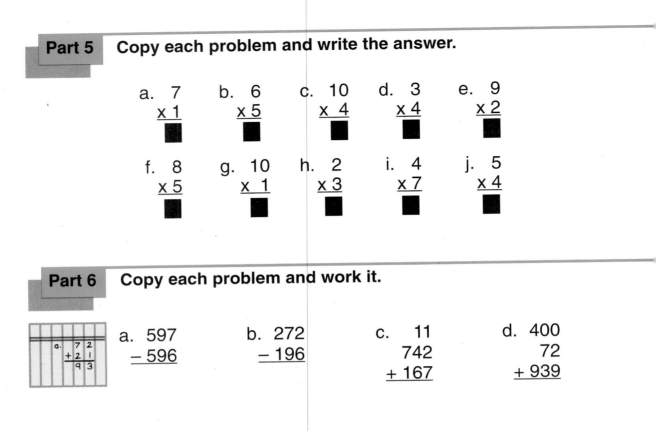

a. 597
 − 596

b. 272
 − 196

c.   11
    742
 + 167

d. 400
     72
 + 939

Do the independent work for lesson 64 in your workbook.

U.S. AIR FORCE

Lesson 65

Part 1   **Make the number family for each problem.**
Write the number problem and figure out the answer.
Remember the unit name.

a. Susan had some books. She gave 95 books away. She
ended up with 69 books. How many books did Susan
start with?

b. Nan had 507 stamps. Joe had 319 fewer stamps than
Nan had. How many stamps did Joe have?

c. A squirrel picked up 185 more sticks than a beaver
picked up. The beaver picked up 246 sticks. How many
sticks did the squirrel pick up?

d. Henry had 192 pieces of wood. He burned some wood.
He ended up with 46 pieces of wood. How many pieces
of wood did Henry burn?

Part 2

- Here's the rule for multiplication number families that have a
small number of 9: The first digit of the big number is 1 less
than the the first small number.

- The first small number of this family is 5:

- So the first digit of the answer is 1 less
than 5. That's 4.

- After you write the first digit of the
answer, you complete the answer.

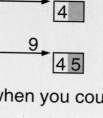

- Remember, the answer is a number you say when you count
by 9. The big number is 45.

- You've figured out missing numbers in a table that's shown twice. You did that by first working the rows in one of the tables and the columns in the other table.

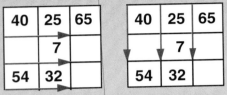

- The number family arrows show that you figure the rows for the first table and the columns for the second table.

40	25	65
	7	
54	32	86

40	25	65
14	7	
54	32	

- When you work the first table, you get a number that you don't get when you work the second table. That number is 86.

- So you copy that number in the second table.

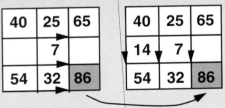

- Now you can figure out the last number in the table because you have two numbers in the column that has a missing number.

65
21
86

Part 4 Copy each problem and work it.

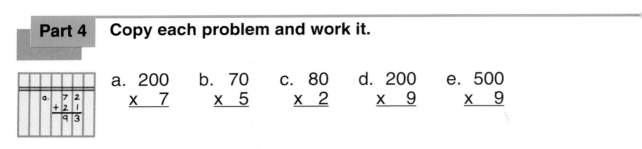

a. 200
   x   7

b. 70
   x   5

c. 80
   x   2

d. 200
   x   9

e. 500
   x   9

Part 5 Write the time shown on each clock.

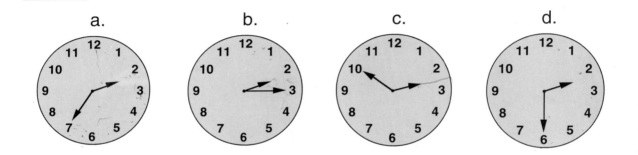

a.

b.

c.

d.

Part 6 Make the number family for each problem.
Write the number problem and figure out the answer.
Remember the unit name.

a. Jim weighs 184 pounds. His horse weighs 808 pounds more than Jim weighs. How much does Jim's horse weigh?

b. A bird stored 314 fewer nuts than a raccoon stored. The raccoon stored 510 nuts. How many nuts did the bird store?

c. You have some. You lose 374. You end up with 381. How many did you start with?

Part 7   Copy each problem and write the answer.

a. $12 - 6 = \blacksquare$ b. $9 - 4 = \blacksquare$ c. $9 - 3 = \blacksquare$ d. $11 - 5 = \blacksquare$

e. $11 - 6 = \blacksquare$ f. $8 - 3 = \blacksquare$ g. $10 - 6 = \blacksquare$ h. $10 - 5 = \blacksquare$

i. $8 - 5 = \blacksquare$ j. $10 - 4 = \blacksquare$ k. $9 - 4 = \blacksquare$ l. $9 - 5 = \blacksquare$

Lesson 66

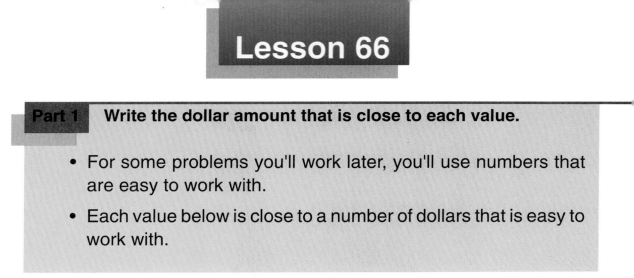

Part 1   Write the dollar amount that is close to each value.

- For some problems you'll work later, you'll use numbers that are easy to work with.
- Each value below is close to a number of dollars that is easy to work with.

Write the dollar amount that is close to each value.

a. $1.90   b. $6.05   c. $4.09   d. $8.91

Part 2   Write the number family for each problem. Then write the number problem and figure out the answer.

a. The mountain is 36 miles farther away than the city is. The mountain is 199 miles away. How far away is the city?

b. A shed took 56 hours to build. A barn took 130 hours longer to build than the shed. How many hours did it take to build the barn?

c. Dan had some trucks. Then he bought 16 more trucks. He ended up with 55 trucks. How many trucks did he start out with?

d. The lake is 41 miles closer than the park. The park is 141 miles away. How far away is the lake?

Copy the table. Figure out the missing numbers.

	30	45
10		35
25		

Independent Work

Part 4 **Write the time for each clock.**

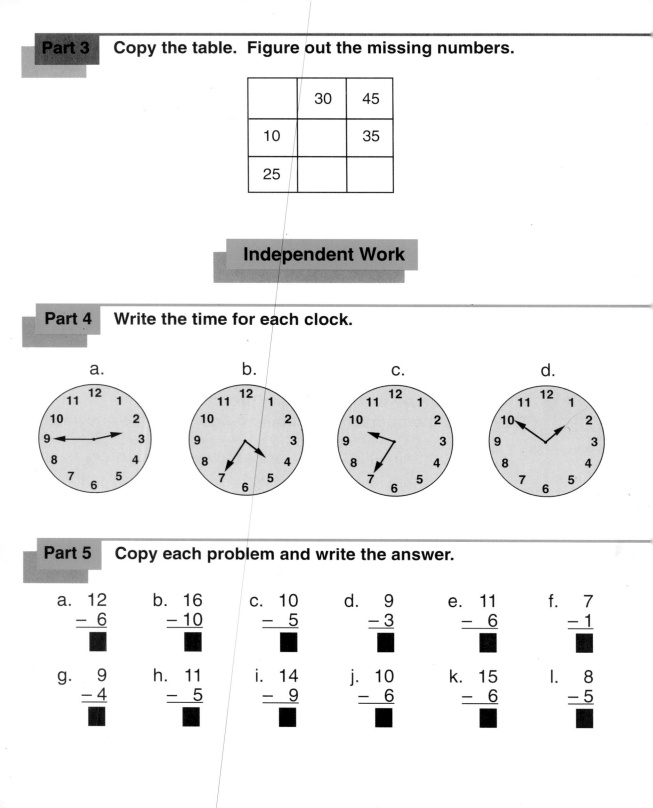

a.   b.   c.   d.

Part 5 **Copy each problem and write the answer.**

a. 12
  − 6
  ■

b. 16
  − 10
  ■

c. 10
  − 5
  ■

d. 9
  − 3
  ■

e. 11
  − 6
  ■

f. 7
  − 1
  ■

g. 9
  − 4
  ■

h. 11
  − 5
  ■

i. 14
  − 9
  ■

j. 10
  − 6
  ■

k. 15
  − 6
  ■

l. 8
  − 5
  ■

Part 6   For each problem, just write the answer.

a.   7 x 2 = ■   b.   3 x 9 = ■   c.   2 x 8 = ■   d.   7 x 10 = ■

e.   5 x 3 = ■   f.   5 x 2 = ■   g.   2 x 6 = ■   h.   4 x 9 = ■

i.   10 x 1 = ■   j.   5 x 5 = ■   k.   9 x 2 = ■   l.   7 x 10 = ■

m.   6 x 5 = ■   n.   4 x 3 = ■   o.   9 x 5 = ■

Part 7   Write the answer to each question.

This table shows the red birds and yellow birds that were seen on Monday and Tuesday.

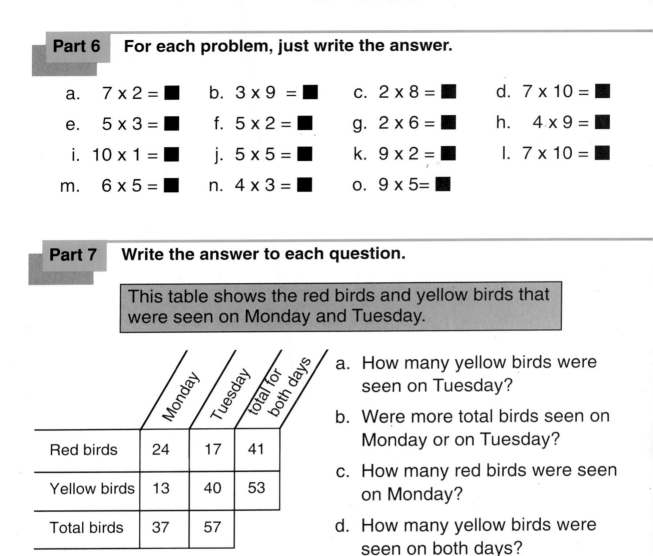

	Monday	Tuesday	total for both days
Red birds	24	17	41
Yellow birds	13	40	53
Total birds	37	57	

a. How many yellow birds were seen on Tuesday?

b. Were more total birds seen on Monday or on Tuesday?

c. How many red birds were seen on Monday?

d. How many yellow birds were seen on both days?

Lesson 67

Part 1

When you write estimation numbers, you write the numbers that are close to each value.

Write the dollar amount that is close to each value.

a. $1.90   b. $6.05   c. $4.09   d. $8.91

e. $12.98   f. $30.91   g. $26.08   h. $15.90

Part 2

Write the number family for each problem. Then write the addition or subtraction problem for each number family and figure out the answer.

a. A cow weighed 196 pounds less than a horse. The cow weighed 741 pounds. How many pounds did the horse weigh?

b. The post was 37 inches longer than the board. The post was 219 inches long. How long was the board?

c. A train started with some tons of concrete. The train unloaded 16 tons of concrete. The train ended up with 986 tons of concrete. How many tons of concrete did the train start with?

d. A toy store had 109 dolls. Then the store bought some more dolls. The store ended up with 525 dolls. How many dolls did the store buy?

Part 3 — Copy the table and figure out the missing numbers.

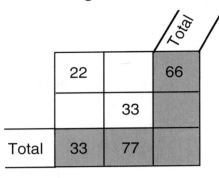

		Total
22		66
	33	
Total	33	77

Independent Work

Part 4 — Write what X equals and what Y equals for each letter.

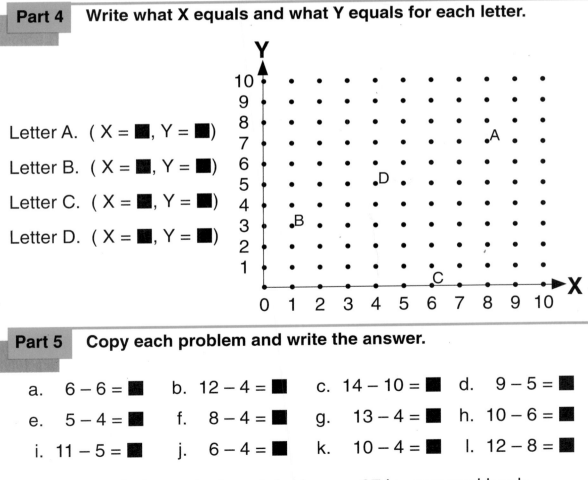

Letter A. (X = ■, Y = ■)

Letter B. (X = ■, Y = ■)

Letter C. (X = ■, Y = ■)

Letter D. (X = ■, Y = ■)

Part 5 — Copy each problem and write the answer.

a. $6 - 6 =$ ■   b. $12 - 4 =$ ■   c. $14 - 10 =$ ■   d. $9 - 5 =$ ■

e. $5 - 4 =$ ■   f. $8 - 4 =$ ■   g. $13 - 4 =$ ■   h. $10 - 6 =$ ■

i. $11 - 5 =$ ■   j. $6 - 4 =$ ■   k. $10 - 4 =$ ■   l. $12 - 8 =$ ■

Do the independent work for lesson 67 in your workbook.

Lesson 68

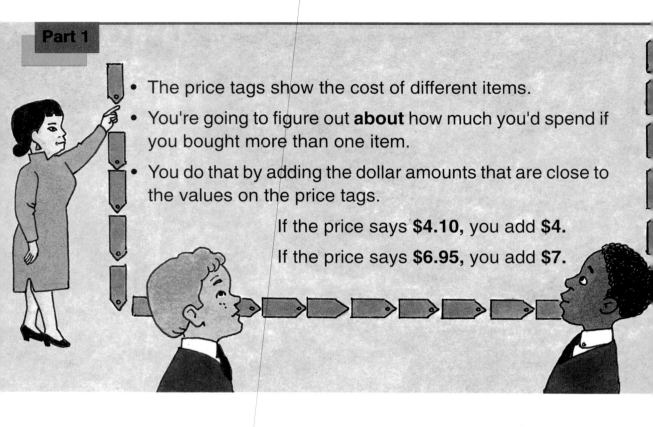

- The price tags show the cost of different items.
- You're going to figure out **about** how much you'd spend if you bought more than one item.
- You do that by adding the dollar amounts that are close to the values on the price tags.

  If the price says **$4.10**, you add **$4**.

  If the price says **$6.95**, you add **$7**.

1	2	3	4	5
$2.92	$3.10	 $ .95	$8.11	 $5.88

a. You want to buy items, 1, 2 and 3. About how much will you spend in all?

b. You want to buy items 2, 3 and 4. About how much will you spend in all?

c. You want to buy items 1, 4 and 5. About how much will you spend in all?

d. You want to buy items 3, 4 and 5. About how much will you spend in all?

- These are multiplication problems that have 9 as a small number.
- So the first digit of the answer is 1 less than the first small number.

Write the answer to each problem.

a. 7 x 9 = ■   b. 4 x 9 = ■   c. 8 x 9 = ■   d. 6 x 9 = ■

e. 9 x 9 = ■   f. 5 x 9 = ■   g. 3 x 9 = ■   h. 2 x 9 = ■

Part 3   **Copy the table. Figure out the missing numbers.**

		66	Total 227
	96		129
Total	257		

Part 4   **Write the number family for each problem. Then write the number problem and figure out the answer.**

a. A woman weighed 142 pounds. She lost some weight. The woman ended up weighing 118 pounds. How many pounds did the woman lose?

b. A plane traveled 718 miles farther than a car traveled. The car traveled 83 miles. How far did the plane travel?

c. Joe had 147 dollars less than Sue. She had 228 dollars. How much money did Joe have?

d. Sue had some money. She earned 406 dollars more. She ended up with 906 dollars. How much money did Sue have to begin with?

Paired Practice

a. 13 − 9 = ■   b. 10 − 4 = ■   c. 12 − 6 = ■   d. 11 − 7 = ■

e. 8 − 4 = ■   f. 9 − 3 = ■   g. 11 − 6 = ■   h. 10 − 5 = ■

i. 12 − 8 = ■   j. 10 − 6 = ■   k. 9 − 6 = ■   l. 9 − 5 = ■

m. 9 − 4 = ■   n. 10 − 6 = ■   o. 8 − 6 = ■   p. 14 − 10 = ■

Independent Work

Part 6   **Copy the problems and figure out the answers.**

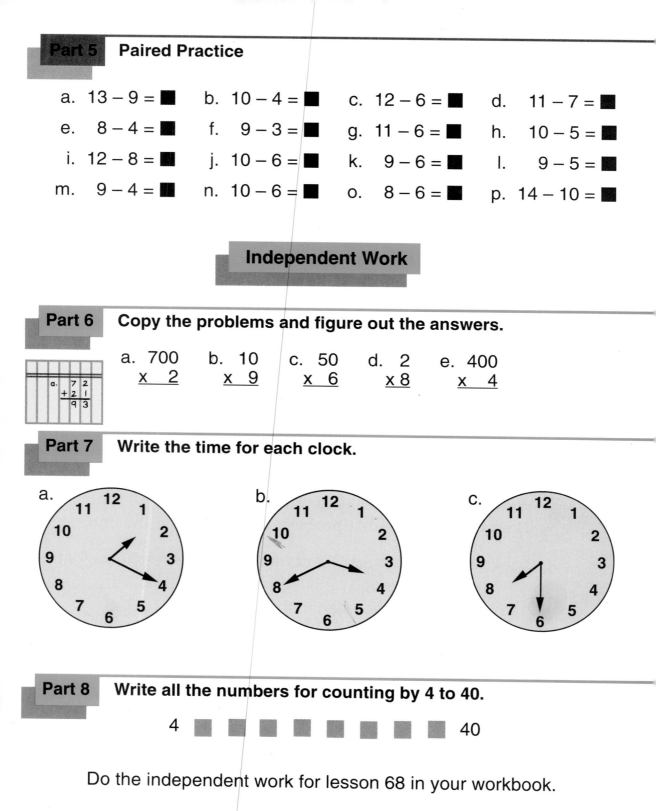

a.	700	b.	10	c.	50	d.	2	e.	400

a. 700   b. 10   c. 50   d. 2   e. 400
    x   2   x 9   x 6   x 8   x   4

Part 7   **Write the time for each clock.**

a.

b.

c.

Part 8   **Write all the numbers for counting by 4 to 40.**

4 ■ ■ ■ ■ ■ ■ ■ ■ 40

Do the independent work for lesson 68 in your workbook.

Lesson 69

Part 1

- The bottom number of a fraction tells how many parts are in each unit.
- The top number tells how many parts are shaded.

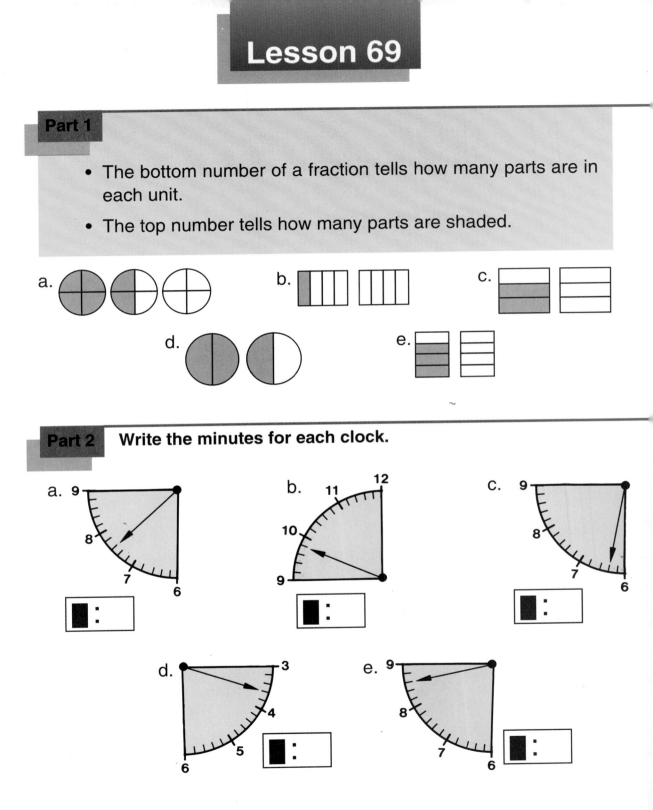

a.

b.

c.

d.

e.

Part 2   Write the minutes for each clock.

a.

b.

c.

d.

e.

Use the dollar amounts that are close to the values on the price tags.

Write the addition problems and answers with a dollar sign.

1	2	3	4	5
$1.90	$3.04	$2.98	$5.09	$1.06

a. You want to buy items, 1, 3 and 4. About how much would you spend?

b. You want to buy items 1 and 5. About how much would you spend?

c. You want to buy items 3, 4 and 5. About how much would you spend?

d. You want to buy items 1, 2 and 4. About how much would you spend?

Part 4   Make the number family for each problem and figure out the answer.

a. The lake is 36 miles closer than the city. The lake is 48 miles away. How far away is the city?

b. Jan worked 47 hours longer than Ted worked. Jan worked 56 hours. How many hours did Ted work?

c. A company had some trucks. Then the company bought 160 more trucks. The company ended up with 255 trucks. How many trucks did the company start with?

d. The mountain is 41 miles closer than the park. The park is 120 miles away. How far away is the mountain?

Paired Practice

a. 9 x 9 = ■   b. 2 x 9 = ■   c. 3 x 9 = ■   d. 4 x 9 = ■

e. 1 x 9 = ■   f. 7 x 9 = ■   g. 8 x 9 = ■   h. 5 x 9 = ■

i. 6 x 9 = ■   j. 3 x 9 = ■   k. 5 x 9 = ■   l. 6 x 9 = ■

m. 8 x 9 = ■   n. 7 x 9 = ■

Independent Work

Part 6   **For each item, write the numbers that could be R.**

a. R ⩾ 73 | 0 100 73 702 | c. R ⩾ 5 | 1 3 7 5 |

b. 1 ⩾ R | 1 10 0 32 |

Part 7   **Copy the problems and figure out the answers.**

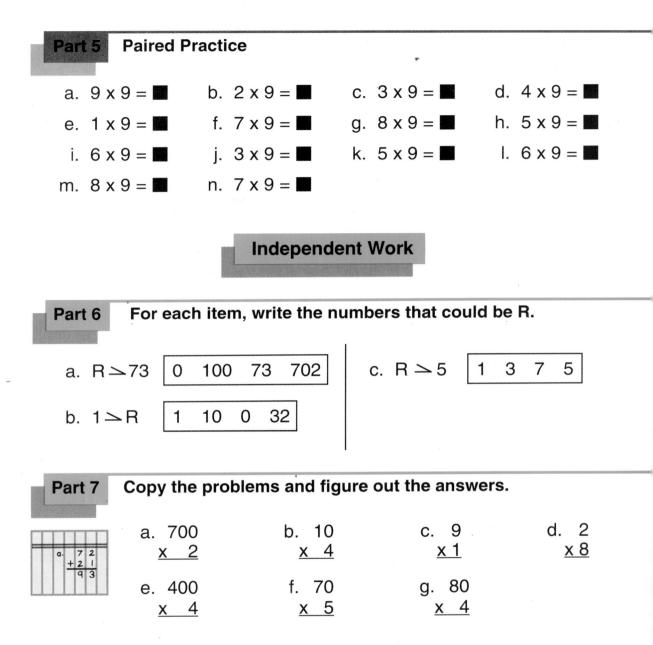

a. 700
  x   2

b.   10
  x   4

c.   9
  x 1

d.   2
  x 8

e. 400
  x   4

f.   70
  x   5

g.   80
  x   4

Do the independent work for lesson 69 in your workbook.

Lesson 70

Part 1

> If a problem compares two things, your number
> family must have two letters.

a. Mary had 36 beans. Then she found some more beans.
   She ended up with 296 beans. How many beans did
   Mary find?

b. An oak tree was 56 feet shorter than a maple tree. The
   oak tree was 165 feet tall. How tall was the maple tree?

c. A redwood tree was 120 feet tall. The redwood tree
   was 78 feet taller than a maple tree. How tall was the
   maple tree?

d. Debbie read some pages in a book. Then she read 36
   more pages. She read 145 pages in all. How many
   pages did she read at first?

e. Dan had $6.90. Then he spent $ .86. How much
   money did he end up with?

Part 2   **Write the minutes for each clock.**

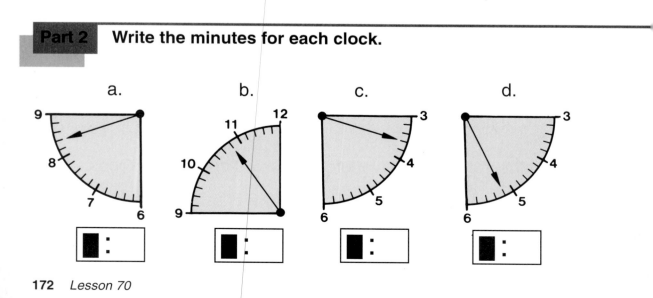

Write the fraction for each item.

a.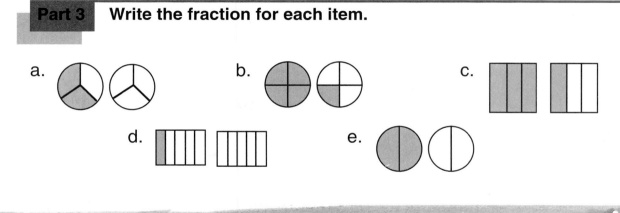

b.

c.

d.

e.

Test 7

Part 7 — Write the answers.

a. $5 \times 9 = \blacksquare$ b. $9 \times 9 = \blacksquare$ c. $8 \times 9 = \blacksquare$ d. $4 \times 9 = \blacksquare$

e. $7 \times 9 = \blacksquare$ f. $3 \times 9 = \blacksquare$ g. $6 \times 9 = \blacksquare$ h. $2 \times 9 = \blacksquare$

Part 8 — For each problem, make the number family and figure out the answer.

a. An oak tree was 56 feet shorter than a maple tree. The oak tree was 65 feet tall. How tall was the maple tree?

b. Dan started out with some dollars in the bank. Then he saved 410 dollars. He ended up with 860 dollars. How many dollars did he start out with?

Part 9 — Copy the problem and write the answer.

a. 700
    × 2

b. 20
    × 9

c. 50
    × 6

d. 500
    × 4

Part 10 — Write the minutes for each clock.

a.

b.

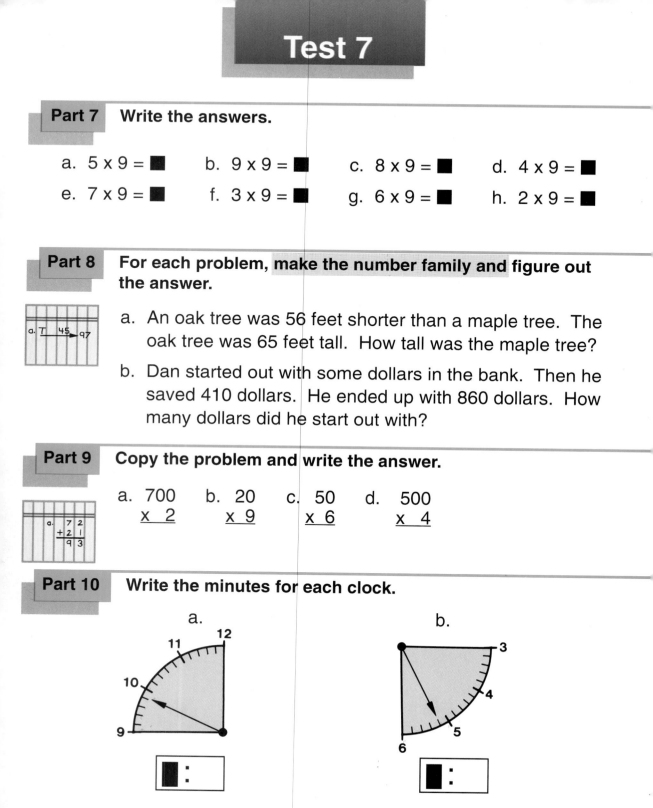

1	2	3	4	5
$1.90	$3.04	$2.98	$5.09	$1.06

a. You want to buy items 3, 4 and 5. **About** how much would you spend?

b. You want to buy items 1, 2 and 4. **About** how much would you spend?

a.
```
  7 2
+ 2 1
-----
  9 3
```

There were 6 people on board. Where are the rest of them?

U.S. COAST GUARD

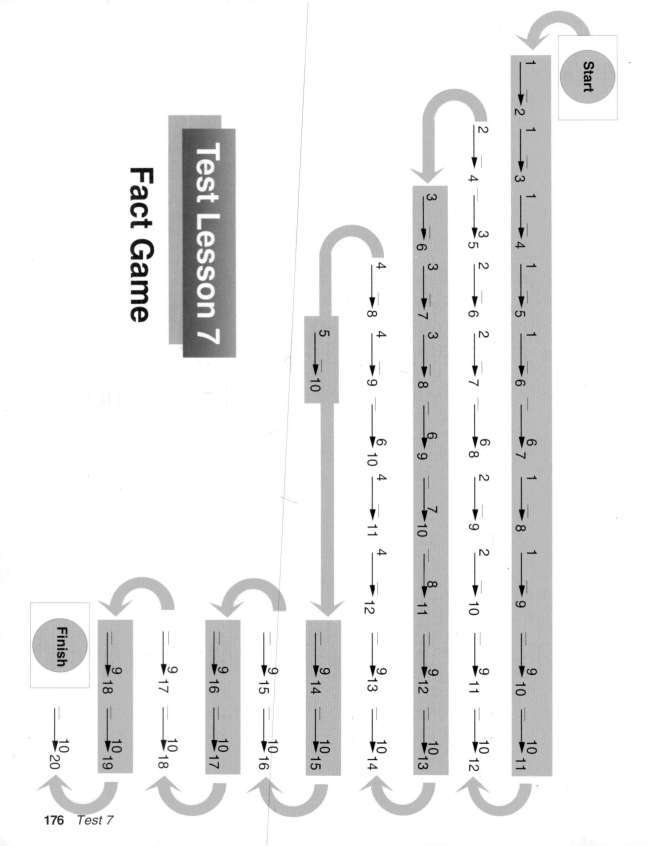

Fact Game

Test Lesson 7

Part 1

- Figure out the minutes for the number just before the minute hand.

- Then keep counting by 1 to the minute hand.

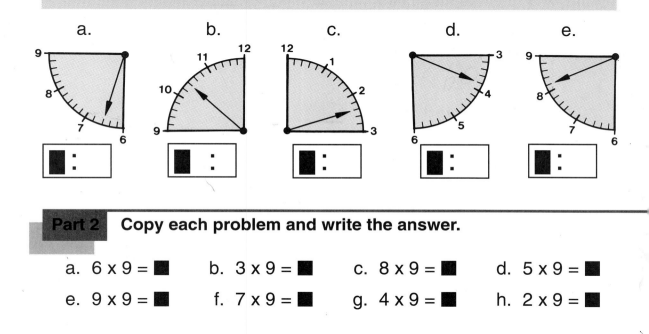

a. b. c. d. e.

Part 2 Copy each problem and write the answer.

a. 6 x 9 = ■

b. 3 x 9 = ■

c. 8 x 9 = ■

d. 5 x 9 = ■

e. 9 x 9 = ■

f. 7 x 9 = ■

g. 4 x 9 = ■

h. 2 x 9 = ■

Part 3

a. Joe worked 137 minutes longer than Debbie. Joe worked 621 minutes. How long did Debbie work?

b. A cow ate 268 pounds less food than a bull. The cow ate 97 pounds of food. How much did the bull eat?

c. There were 86 more clouds on Tuesday than Wednesday. On Wednesday there were 49 clouds. How many clouds were there on Tuesday?

d. A rabbit was 15 yards farther away than a turtle. The rabbit was 63 yards away. How far away was the turtle?

e. An elm tree was 45 feet shorter than a maple. The maple was 112 feet tall. How tall was the elm?

Independent Work

Part 4 **Copy the problems and write the answers.**

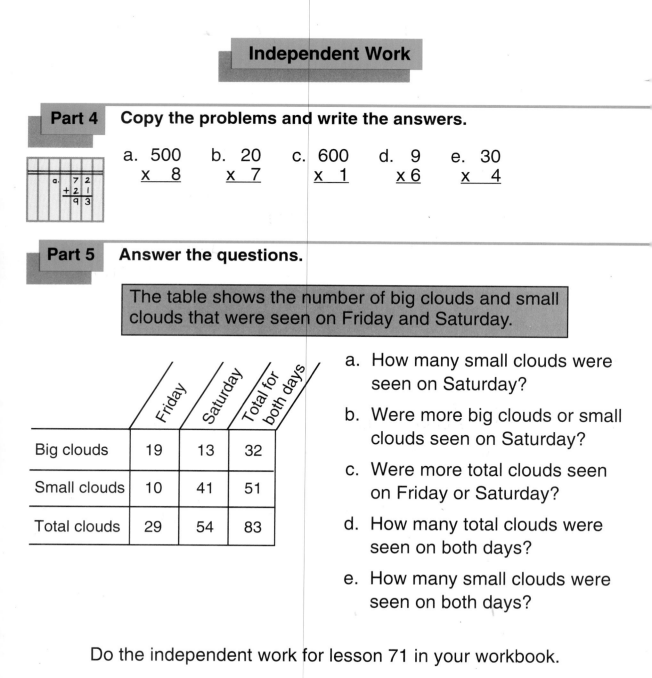

a. 500   b. 20   c. 600   d. 9   e. 30
   x   8   x 7   x 1   x 6   x   4

Part 5 **Answer the questions.**

The table shows the number of big clouds and small clouds that were seen on Friday and Saturday.

	Friday	Saturday	Total for both days
Big clouds	19	13	32
Small clouds	10	41	51
Total clouds	29	54	83

a. How many small clouds were seen on Saturday?

b. Were more big clouds or small clouds seen on Saturday?

c. Were more total clouds seen on Friday or Saturday?

d. How many total clouds were seen on both days?

e. How many small clouds were seen on both days?

Do the independent work for lesson 71 in your workbook.

Lesson 72

Part 1 Write the answer to each problem. Do not copy the problems.

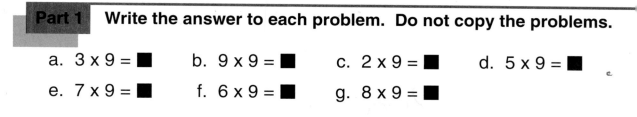

a. 3 x 9 = ■   b. 9 x 9 = ■   c. 2 x 9 = ■   d. 5 x 9 = ■

e. 7 x 9 = ■   f. 6 x 9 = ■   g. 8 x 9 = ■

Part 2

a. Kim had 67 pieces of wood. She bought some more wood. She ended up with 499 pieces of wood. How many pieces of wood did she buy?

b. Kim had some wood. She burned up 67 pieces of wood. She ended up with 499 pieces of wood. How much wood did she start out with?

c. Kim had 499 pieces of wood. She gave away some wood. She ended up with 76 pieces of wood. How many pieces of wood did she give away?

d. Kim had some wood. She bought 167 pieces of wood. She ended up with 499 pieces of wood. How many pieces did she start out with?

Part 3 Write answers to all the problems. Don't copy the problems. Just write the answers.

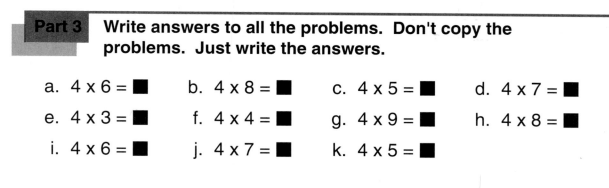

a. 4 x 6 = ■   b. 4 x 8 = ■   c. 4 x 5 = ■   d. 4 x 7 = ■

e. 4 x 3 = ■   f. 4 x 4 = ■   g. 4 x 9 = ■   h. 4 x 8 = ■

i. 4 x 6 = ■   j. 4 x 7 = ■   k. 4 x 5 = ■

- Here's how you figure out the missing numbers in a table.
- First, you work all the **rows** that have two numbers.
- Then you work all the **columns** that have two numbers.

The table shows the clouds that were seen on Wednesday and Thursday.

	Wednesday	Thursday	Total for both days
Big clouds		84	
Small clouds	92	115	
Total clouds		199	321

a. On which day were more total clouds seen?

b. Were there more small clouds seen on Wednesday or on Thursday?

c. Were there more big clouds seen on Wednesday or on Thursday?

d. How many small clouds were seen on Thursday?

e. How many big clouds were seen on Wednesday?

Part 5 **Paired Practice**

a. $16 - 8 = \blacksquare$ b. $16 - 10 = \blacksquare$ c. $6 - 3 = \blacksquare$ d. $6 - 5 = \blacksquare$

e. $6 - 2 = \blacksquare$ f. $12 - 6 = \blacksquare$ g. $12 - 10 = \blacksquare$ h. $4 - 0 = \blacksquare$

i. $4 - 4 = \blacksquare$ j. $10 - 10 = \blacksquare$ k. $10 - 5 = \blacksquare$ l. $18 - 10 = \blacksquare$

m. $18 - 9 = \blacksquare$ n. $14 - 7 = \blacksquare$ o. $7 - 7 = \blacksquare$

Part 6

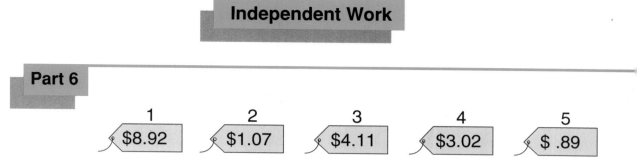

1	2	3	4	5
$8.92	$1.07	$4.11	$3.02	$ .89

Use estimation. Write dollar amounts that are close to the values on the price tags.
Write the addition problems and the answers with a dollar sign.

a. You buy items 1, 3 and 5. About how much do you spend?

b. You buy items 2, 3 and 5. About how much do you spend?

c. You buy items 1, 2 and 4. About how much do you spend?

d. You buy items 1, 4 and 5. About how much do you spend?

Use the values on the price tags and add to see exactly how much you spend.
Write the addition problems and the answer with a dollar sign.

a. You buy items 1, 3 and 5. How much do you spend?

b. You buy items 2, 3 and 5. How much do you spend?

c. You buy items 1, 2 and 4. How much do you spend?

d. You buy items 1, 4 and 5. How much do you spend?

Do the independent work for lesson 72 in your workbook.

Lesson 73

Part 1

- These are multiplication problems based on number families that have 4 as a small number.

- In some problems the 4 is first, and in other problems the 4 is second.

- But if there is a 4, you'll know how to work the problem if you think of the number family.

a. $4 \times 6 = \blacksquare$ b. $4 \times 9 = \blacksquare$ c. $4 \times 7 = \blacksquare$ d. $4 \times 8 = \blacksquare$

e. $3 \times 4 = \blacksquare$ f. $6 \times 4 = \blacksquare$ g. $8 \times 4 = \blacksquare$ h. $10 \times 4 = \blacksquare$

i. $4 \times 7 = \blacksquare$ j. $5 \times 4 = \blacksquare$ k. $9 \times 4 = \blacksquare$ l. $4 \times 3 = \blacksquare$

Part 2   **Copy the table. Figure out the missing numbers. Then write the answers to the questions.**

This table shows the number of red birds and blue birds that were seen on Monday and Friday.

	Monday	Friday	Total for both days
Red birds	19		82
Blue birds	33		
Total birds	52	121	

a. On Monday, were more red birds or blue birds seen?

b. On Friday, were more red birds or blue birds seen?

c. How many total birds were seen on both days?

d. How many total birds were seen on Friday?

Part 3 Make the number family for each problem.
Write the number problem and figure out the answer.
Remember the unit name.

a. A boy had some flowers. He grew 136 more flowers.
He ended up with 222 flowers. How many flowers did
the boy start out with?

b. A cow ate 67 pounds more alfalfa than a horse ate. The
cow ate 96 pounds of alfalfa. How much alfalfa did the
horse eat?

c. Joe weighed 238 pounds. He lost some weight. He
ended up weighing 150 pounds. How much weight did
Joe lose?

d. A lion weighed 238 pounds less than a bear weighed.
The lion weighed 565 pounds. How much did the bear
weigh?

Part 4 Copy each problem
and work it.

a. 700   b. 80   c. 5
  x   4   x 2   x 4

Part 5 Copy each problem
and work it.

a. 307   b.   8
     10   691
  + 894   + 34

Use estimation. Write dollar amounts that are close to the
values on the price tags. Then write the addition problems
and the answers with a dollar sign.

1	2	3	4	5
$1.98	$2.89	$3.06	$9.92	$ .94

a. A boy buys items 2, 4 and 5. About how much did he spend?

b. A boy buys items 1, 2 and 3. About how much did he spend?

c. A boy buys items 2, 3 and 5. About how much did he spend?

d. A boy buys items 1, 3 and 4. About how much did he spend?

Part 1 Copy the table. Figure out the missing numbers. Then write answers to the questions.

> This table shows the number of old cars and new cars that were sold on Sunday and Monday.

	Sunday	Monday	Total for both days
Old cars			69
New cars	44	46	
Total cars		75	159

a. On Sunday, were there more old cars or new cars sold?

b. How many total cars were sold on both days?

c. On Monday, were there more old cars or new cars sold?

d. How many total cars were sold on Monday?

This band needs 8 more horns.

- You can use numbers to show the values in a column. Or you can use pictures to show the same values.

- Here are numbers that show how many pounds of berries were picked on Monday and on Tuesday.

Day	Pounds of Berries Picked on 2 days
Monday	4
Tuesday	2
Total for both days	6

- Here is a picture graph that shows the same values. Each berry represents one pound of berries.

Day	Pounds of Berries Picked on 2 Days
Monday	🫐🫐🫐🫐
Tuesday	🫐🫐
Total for both days	🫐🫐🫐🫐🫐🫐

- Here's another kind of graph. It's called a bar graph. The bars on the graph start in the same place.

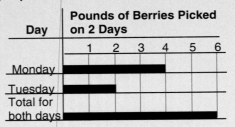

- The bar for Monday is 4 units long. The end of that bar is at the line for 4 units. Each unit represents one pound of berries.

- The bar for Tuesday is 2 units long.

- The bar for the total is 6 units long.

A. Inches of Rainfall in 2 Cities

City	Inches
River City	🌢🌢🌢🌢🌢🌢🌢🌢🌢
Mountain City	🌢🌢🌢🌢🌢🌢
Total for both cities	🌢🌢🌢🌢🌢🌢🌢 🌢🌢🌢🌢🌢🌢🌢

1. Did more inches of rain fall in River City or in Mountain City?

2. How many inches of rain fell in River City?

3. How many inches fell in both cities?

B. Trucks Travelling on 2 Streets

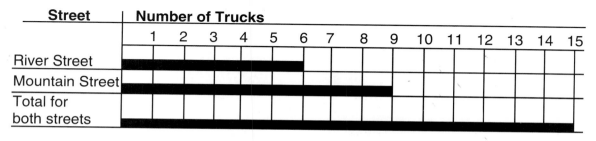

Street	Number of Trucks														
	1	2	3	4	5	6	7	8	9	10	11	12	13	14	15
River Street															
Mountain Street															
Total for both streets															

1. Did more trucks go down River Street or Mountain Street?

2. How many total trucks went down both streets?

3. How many trucks went down Mountain Street?

C. Trees Planted in 2 Parks

Parks	Trees
River Park	🌲🌲🌲
Mountain Park	🌲🌲🌲🌲
Total for both Parks	🌲🌲🌲🌲🌲🌲🌲

1. How many trees were planted in both parks?

2. In which park were more trees planted?

3. How many trees were planted in River Park?

Write the answers to all the subtraction problems.

a. $8 - 3 = \blacksquare$ b. $10 - 3 = \blacksquare$ c. $7 - 3 = \blacksquare$ d. $11 - 3 = \blacksquare$

e. $6 - 3 = \blacksquare$ f. $12 - 3 = \blacksquare$ g. $10 - 3 = \blacksquare$ h. $13 - 3 = \blacksquare$

i. $11 - 3 = \blacksquare$

Independent Work

Part 4 **Write the minutes for each clock.**

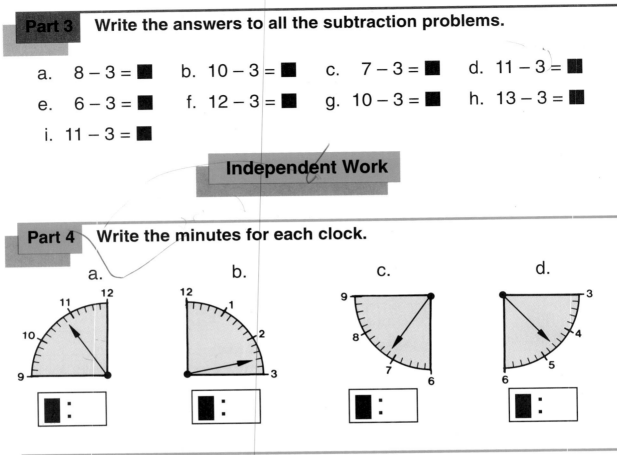

a.   b.   c.   d.

Part 5 **For each problem, write the number family. Then figure out the answer. Remember the unit name.**

a. The truck held 37 fewer boxes than the van. The van held 171 boxes. How many boxes did the truck hold?

b. Carlos bought 131 marbles. He gave some to his sister. Now he has 41 marbles. How many marbles did he give to his sister?

c. There were 141 more women than men at the beach. There were 765 women at the beach. How many men were at the beach?

Do the independent work for lesson 74 in your workbook.

Lesson 75

Part 1

When you read a fraction, you start with the top number.
This fraction is three-fourths: $\frac{3}{4}$

a. $\frac{5}{4}$ b. $\frac{7}{4}$ c. $\frac{7}{8}$ d. $\frac{1}{7}$ e. $\frac{5}{7}$ f. $\frac{2}{9}$ g. $\frac{6}{10}$

Part 2

A. Inches of Rainfall in 2 Cities During May

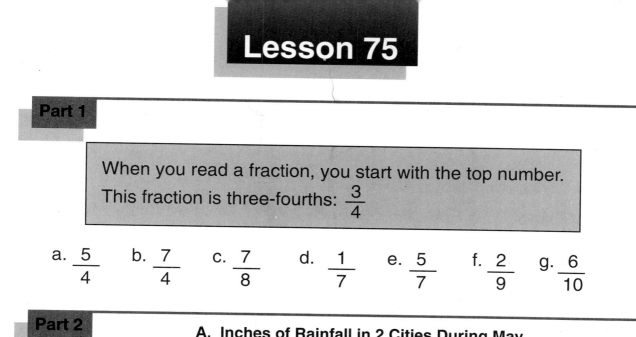

City	Inches									
	$\frac{1}{4}$	$\frac{2}{4}$	$\frac{3}{4}$	$\frac{4}{4}$	$\frac{5}{4}$	$\frac{6}{4}$	$\frac{7}{4}$	$\frac{8}{4}$	$\frac{9}{4}$	$\frac{10}{4}$
Flatville										
Mudville										
Total										

1. How many inches of rain fell in Flatville?
2. How many inches of rain fell in both cities?

B. Trucks Stopping in 2 Cities on Monday

City	Number of Trucks									
	1	2	3	4	5	6	7	8	9	10
Flatville										
Mudville										
Total										

1. How many trucks stopped in Flatville?
2. How many trucks stopped in both cities?

C. Length of Main Street in 2 Cities

City	Miles									
	$\frac{1}{10}$	$\frac{2}{10}$	$\frac{3}{10}$	$\frac{4}{10}$	$\frac{5}{10}$	$\frac{6}{10}$	$\frac{7}{10}$	$\frac{8}{10}$	$\frac{9}{10}$	$\frac{10}{10}$
Flatville										
Mudville										
Total										

1. How many miles long is the Main Street in Mudville?

2. How many miles long is the Main Street in Flatville?

Write the answers to all the problems.

a. 7 − 3 = ■   b. 10 − 3 = ■   c. 12 − 3 = ■   d. 8 − 3 = ■

e. 11 − 3 = ■   f. 7 − 3 = ■   g. 9 − 3 = ■   h. 10 − 3 = ■

i. 6 − 3 = ■   j. 9 − 3 = ■   k. 8 − 3 = ■   l. 11 − 3 = ■

m. 13 − 3 = ■

Part 4 Copy the table. Figure out the missing numbers. Then write answers to the questions.

This table shows the number of small rocks and big rocks that were seen in a valley and on a hill.

	Valley	Hill	Total for both places
Small rocks			145
Big rocks	29	36	
Total rocks	59		210

a. How many big rocks were seen in both places?

b. Were more small rocks seen in the valley or on the hill?

c. How many total rocks were seen in both places?

d. Were more small rocks or big rocks seen on the hill?

Part 5   **Use estimation. Write dollar amounts that are close to the values on the price tags. Then write the addition problems and the answers with a dollar sign.**

1	2	3	4	5
$ .99	$8.95	$2.06	$2.96	$7.04

a. A woman buys items 1, 2 and 4. About how much does she spend?

b. A boy buys items 3, 4 and 5. About how much does he spend?

c. A man buys items 1, 3 and 4. About how much does he spend?

d. A girl buys items 1, 2 and 5. About how much does she spend?

Do the independent work for lesson 75 in your workbook.

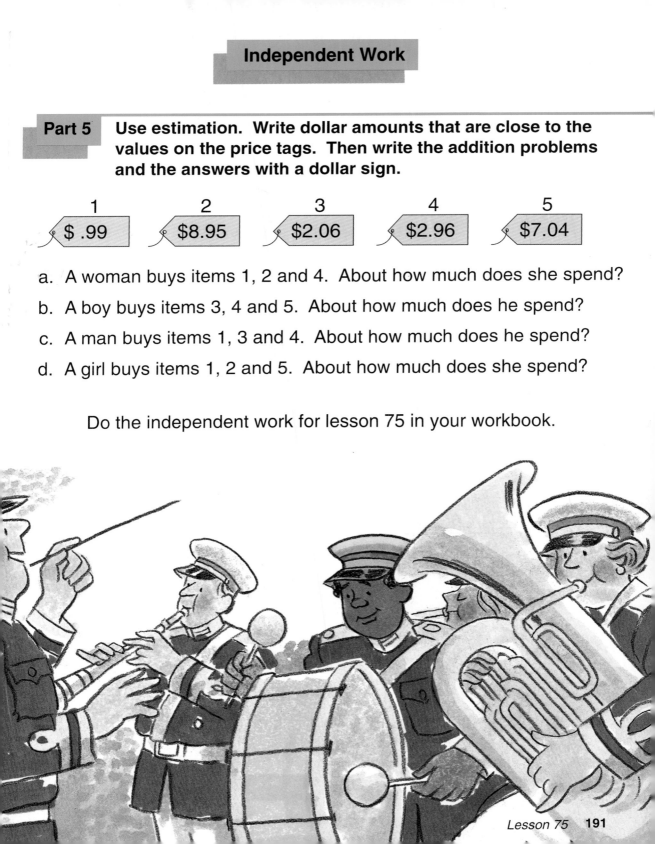

Lesson 76

Part 1

Each graph shows things that happened in the months of June, July and August.

A. Number of Cloudy Days

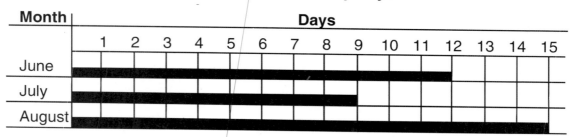

Month	Days 1	2	3	4	5	6	7	8	9	10	11	12	13	14	15
June															
July															
August															

1. How many cloudy days were there in August?
2. How many cloudy days were there in June?

B. Inches of Rainfall

Month	Inches (1 raindrop = 1 inch)
June	🌢🌢🌢🌢🌢🌢🌢
July	🌢🌢🌢🌢🌢🌢
August	🌢🌢

1. How many inches of rain fell in July?
2. How many inches of rain fell in August?
3. In which month did the most inches of rain fall?

Part 2

- For some problems you'll use estimation numbers to figure out **about** how much the person spends.

- For some problems, you'll use the exact numbers to figure out **exactly** how much the person spends.

Write the addition problems and answers with a dollar sign.

1	2	3	4	5
$2.98	$1.36	$ .92	$6.08	$4.10

a. A boy buys items 1, 3 and 4. **About** how much money does he spend?

b. A boy buys items 1, 3 and 4. **Exactly** how much money does he spend?

c. A man buys items 3, 4 and 5. **About** how much money does he spend?

d. A woman buys items 3, 4 and 5. **Exactly** how much money does she spend?

Paired Practice

a. $3 \times 4 = $ ■ b. $5 \times 4 = $ ■ c. $7 \times 4 = $ ■ d. $9 \times 4 = $ ■

e. $8 \times 4 = $ ■ f. $6 \times 4 = $ ■ g. $4 \times 4 = $ ■ h. $10 \times 4 = $ ■

i. $4 \times 9 = $ ■ j. $4 \times 6 = $ ■ k. $4 \times 8 = $ ■ l. $4 \times 3 = $ ■

m. $4 \times 10 = $ ■ n. $4 \times 7 = $ ■ o. $4 \times 2 = $ ■ p. $4 \times 5 = $ ■

Independent Work

Part 4 **Make a number family for each problem.**
Then figure out the answer. Remember the unit name.

a. A dog had some fleas. Then 59 left the dog. The dog ended up with 451 fleas. How many fleas did the dog start out with?

b. Joe worked 99 hours longer than Betty. Joe worked 148 hours. How long did Betty work?

c. Jim had 194 chickens. He bought some more chickens. He ended up with 308 chickens. How many chickens did Jim buy?

Part 5 **Write the minutes for each clock.**

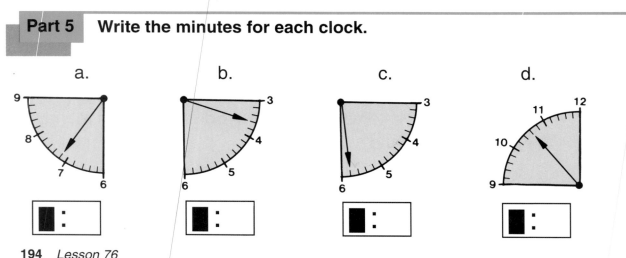

a. b. c. d.

Lesson 77

Part 1

Here's more about fractions:

- When the bottom number is **2**, the fraction is **halves**.
- When the bottom number is **3**, the fraction is **thirds**.
- When the bottom number is **5**, the fraction is **fifths**.

a. $\dfrac{5}{2}$ b. $\dfrac{1}{2}$ c. $\dfrac{5}{3}$ d. $\dfrac{2}{5}$ e. $\dfrac{7}{2}$ f. $\dfrac{6}{3}$

Part 2   Write the multiplication fact that begins with the missing number.

a. 2 ⟶ 10   b. 2 ⟶ 8   c. 2 ⟶ 20   d. 7 ⟶ 70

e. 2 ⟶ 16   f. 2 ⟶ 12   g. 2 ⟶ 18

Part 3   Write answers to all the problems.

a. $\begin{array}{r} 10 \\ -\ 7 \\ \hline \end{array}$ b. $\begin{array}{r} 6 \\ -\ 3 \\ \hline \end{array}$ c. $\begin{array}{r} 8 \\ -\ 3 \\ \hline \end{array}$ d. $\begin{array}{r} 10 \\ -\ 3 \\ \hline \end{array}$ e. $\begin{array}{r} 11 \\ -\ 8 \\ \hline \end{array}$ f. $\begin{array}{r} 13 \\ -\ 3 \\ \hline \end{array}$

g. $\begin{array}{r} 11 \\ -\ 3 \\ \hline \end{array}$ h. $\begin{array}{r} 10 \\ -\ 7 \\ \hline \end{array}$ i. $\begin{array}{r} 12 \\ -\ 8 \\ \hline \end{array}$ j. $\begin{array}{r} 13 \\ -\ 9 \\ \hline \end{array}$ k. $\begin{array}{r} 11 \\ -\ 8 \\ \hline \end{array}$ l. $\begin{array}{r} 7 \\ -\ 3 \\ \hline \end{array}$

Part 4 Use dollar amounts that are close to the values on the price tags to figure **about** how much.
Use the values on the price tag to figure **exactly** how much.

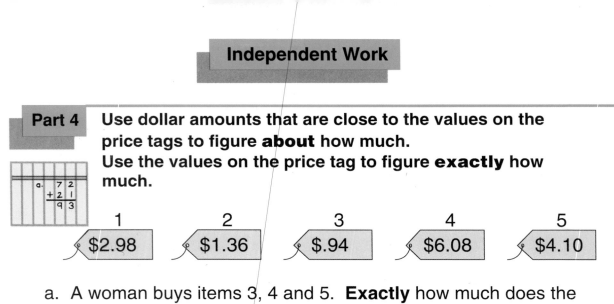

1	2	3	4	5
$2.98	$1.36	$.94	$6.08	$4.10

a. A woman buys items 3, 4 and 5. **Exactly** how much does the woman have left?

b. A boy buys items 2, 3 and 5. **About** how much does the boy spend?

c. A girl buys items 1, 3 and 4. **Exactly** how much does the girl spend?

d. A girl buys items 3,4 and 5. **About** how much does she spend?

Part 5 Make a number family for each problem.
Write the column problem and figure out the answer.
Remember the unit name.

a. Joe has $56 more than Debbie. Joe has $115. How much does Debbie have?

b. Henry had some stamps. He gave away 130 stamps. He ended up with 454 stamps. How many stamps did he start out with?

c. A frog ate 137 fewer bugs than a sparrow ate. The frog ate 64 bugs. How many bugs did the sparrow eat?

Lesson 78

Part 1 Read the fractions.

a. $\dfrac{3}{2}$ b. $\dfrac{5}{2}$ c. $\dfrac{7}{5}$ d. $\dfrac{2}{3}$ e. $\dfrac{1}{5}$ f. $\dfrac{2}{2}$

Part 2 Paired Practice

a. $8 - 3 = $ ■   b. $12 - 6 = $ ■   c. $6 - 3 = $ ■   d. $14 - 7 = $ ■

e. $8 - 5 = $ ■   f. $7 - 3 = $ ■   g. $10 - 7 = $ ■   h. $11 - 3 = $ ■

i. $11 - 10 = $ ■   j. $8 - 4 = $ ■   k. $11 - 8 = $ ■   l. $12 - 3 = $ ■

m. $11 - 9 = $ ■   n. $10 - 5 = $ ■   o. $7 - 4 = $ ■   p. $10 - 3 = $ ■

Part 3

- The bottom number and the top number in $\dfrac{3}{3}$ are the same.
- The fraction equals 1.

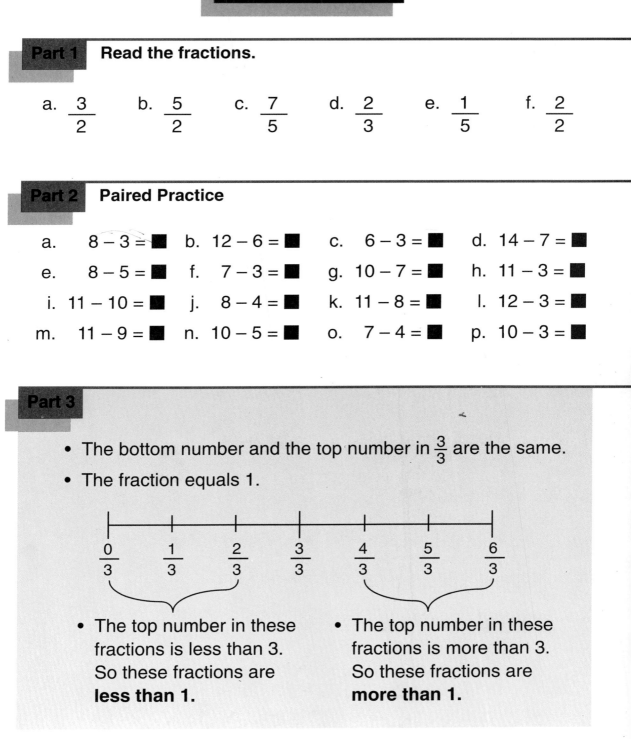

$\dfrac{0}{3} \quad \dfrac{1}{3} \quad \dfrac{2}{3} \quad \dfrac{3}{3} \quad \dfrac{4}{3} \quad \dfrac{5}{3} \quad \dfrac{6}{3}$

- The top number in these fractions is less than 3. So these fractions are **less than 1.**

- The top number in these fractions is more than 3. So these fractions are **more than 1.**

Part 4

Make a number family for each problem.
Write the column problem and figure out the answer.
Remember the unit name.

a. Doris planted some trees on Tuesday. Then she planted 34 trees on Wednesday. She planted 97 trees in all. How many did she plant on Tuesday?

b. Don planted 45 more trees than Fran planted. Don planted 124 trees. How many trees did Fran plant?

c. The goat weighed 32 pounds less than the sheep. The goat weighed 135 pounds. How much did the sheep weigh?

Part 5

Copy each problem and work it.

a. $\begin{array}{r} 600 \\ \times\ \ 4 \\ \hline \end{array}$

b. $\begin{array}{r} 90 \\ \times\ \ 4 \\ \hline \end{array}$

c. $\begin{array}{r} 300 \\ \times\ \ 4 \\ \hline \end{array}$

Part 6

Write the minutes for each clock.

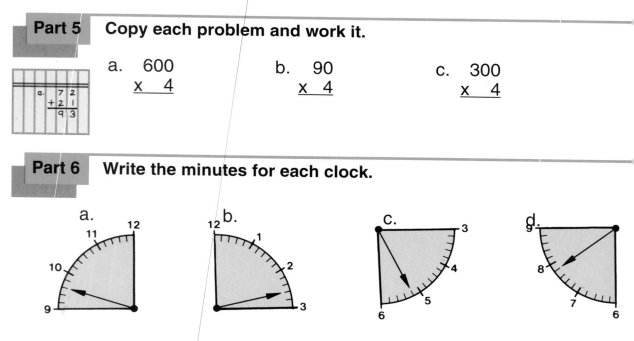

a.

b.

c.

d.

Use dollar amounts that are close to the values on the price tag to figure **about** how much.
Use the value on the price tags to figure **exactly** how much.

a.
	7	2
+	2	1
	9	3

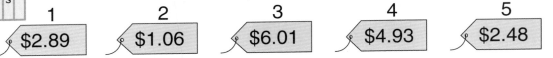

1	2	3	4	5
$2.89	$1.06	$6.01	$4.93	$2.48

a. A woman buys items 2, 3 and 5. **Exactly** how much does she spend?

b. A girl buys items 1, 2 and 4. **About** how much does the girl spend?

c. A boy buys items 1, 2 and 3. **About** how much does the boy spend?

We can sing for 5 minutes more.

Part 1

- You're going to write the time shown on each clock.
- The minute hand is not pointing to a number on the clock.
- Remember how to figure out the minutes. Count by fives to the number just **before** the minute hand.
- Then count by ones to the minute hand.

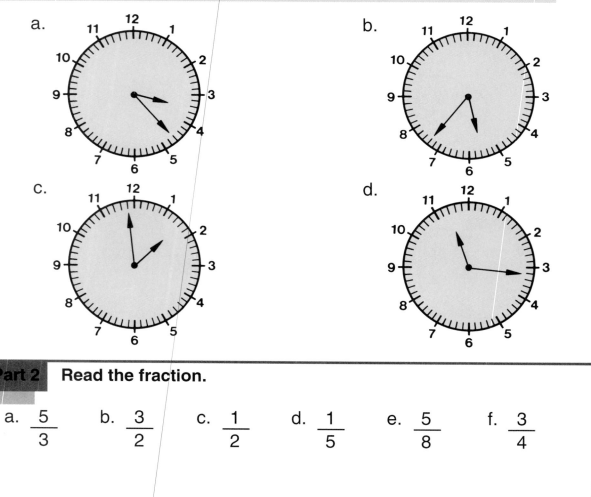

a.

b.

c.

d.

Part 2 Read the fraction.

a. $\dfrac{5}{3}$ b. $\dfrac{3}{2}$ c. $\dfrac{1}{2}$ d. $\dfrac{1}{5}$ e. $\dfrac{5}{8}$ f. $\dfrac{3}{4}$

Write answers to all the problems.

a. $14 - 5$ b. $13 - 5$ c. $10 - 5$ d. $15 - 5$ e. $11 - 5$

f. $12 - 5$ g. $13 - 5$ h. $14 - 5$ i. $15 - 5$ j. $12 - 5$

Independent Work

Part 4 Make a number family for each problem. Then write the addition or subtraction problem for each number family and figure out the answer. Remember the unit name.

a. T 45 97

a. In the morning, Fred read 24 pages. In the afternoon, he read some more pages. He read a total of 71 pages that day. How many pages did he read in the afternoon?

b. Fran read 29 fewer pages than Greg read. Fran read 71 pages. How many pages did Greg read?

c. Joe had some bird houses. Joe sold 72 bird houses. He ended up with 121 bird houses. How many bird houses did Joe start with?

d. Joe had 245 more nails than Debbie. Joe had 340 nails. How many nails did Debbie have?

Part 5 Copy each problem and work it.

a.
+
7 2
2 1
9 3

a. 7×4 b. 400×2 c. 60×4 d. 90×4 e. 500×3

Part 1

- You can use estimation to work these problems the same way you work estimation problems with money.

- Here's a problem:
$$\begin{array}{r} 402 \\ + 398 \end{array}$$

- The first number is close to 400.

- The second number is close to 400.

- So we just add 400 and 400 to get an estimation of the answer.

$$\begin{array}{r} 400 \\ + 400 \\ \hline 800 \end{array}$$

a.	$\begin{array}{r}72\\+21\\\hline 93\end{array}$				

a. $\begin{array}{r} 402 \\ + 398 \end{array}$
b. $\begin{array}{r} 206 \\ + 691 \end{array}$
c. $\begin{array}{r} 597 \\ - 399 \end{array}$
d. $\begin{array}{r} 699 \\ - 204 \end{array}$
e. $\begin{array}{r} 406 \\ + 292 \end{array}$

Part 2 **Copy the table.**
Then figure out the missing numbers.

	Elm Street	Oak Street	Total for both streets
Blue birds			
Black birds	32		
Total birds			

Fact 1: There were 29 blue birds.

Fact 2: The total birds seen on Elm Street was 38.

Fact 3: There were 37 black birds seen on Oak Street.

Fact 4: There were 6 blue birds seen on Elm Street.

Write the answer to each problem.

a. 10
    − 5

b. 12
    − 5

c. 13
    − 5

d. 15
    − 5

e. 14
    − 5

f. 11
    − 5

g. 12
    − 5

h. 13
    − 5

i. 11
    − 5

Part 4   **Write the time shown on each clock.**

Independent Work

Part 5   Make a number family for each problem. Then write the
number problem and the answer. Remember the unit name.

a. Jane had some nails. She found 36 nails. She ended up
with 324 nails. How many nails did she start out with?

b. 19 inches of snow fell in January. Then some more fell in
February. If 46 inches fell in both months, how many
inches of snow fell in February?

c. Joe had 368 toys. He gave some toys away. He ended
up with 99 toys. How many toys did Joe give away?

d. A goat ate 384 more ounces of food than a sheep ate.
The goat ate 432 ounces of food. How many ounces of
food did the sheep eat?

Part 6   Use dollars amounts that are close to the values on the
price tags to figure out **about** how much.
Use the values on the price tags to figure out **exactly** how
much.

1   $ .89   2 $4.03   3 $8.11   4 $1.91   5 $5.96

a. A girl buys items 1, 3 and 4. **Exactly** how much does she
spend?

b. A boy buys items 3, 4 and 5. **About** how much does he spend?

c. A man buys items 1 and 5. **About** how much does he spend?

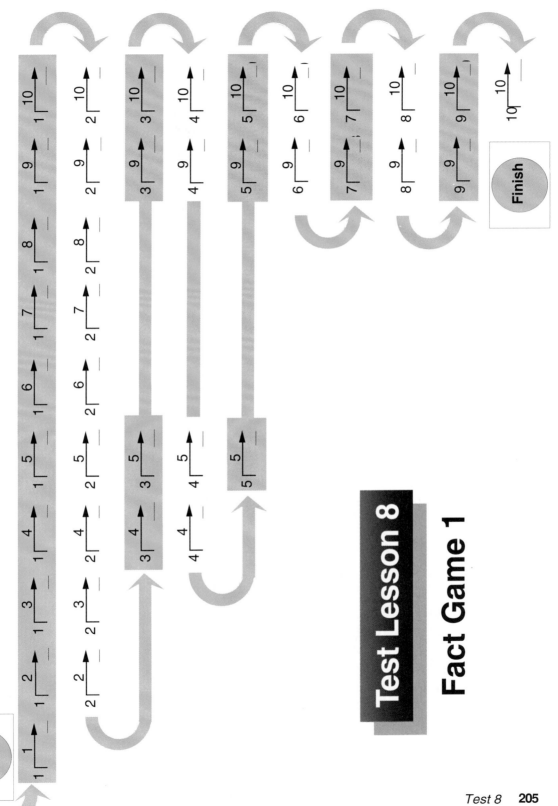

Test Lesson 8

Fact Game 1

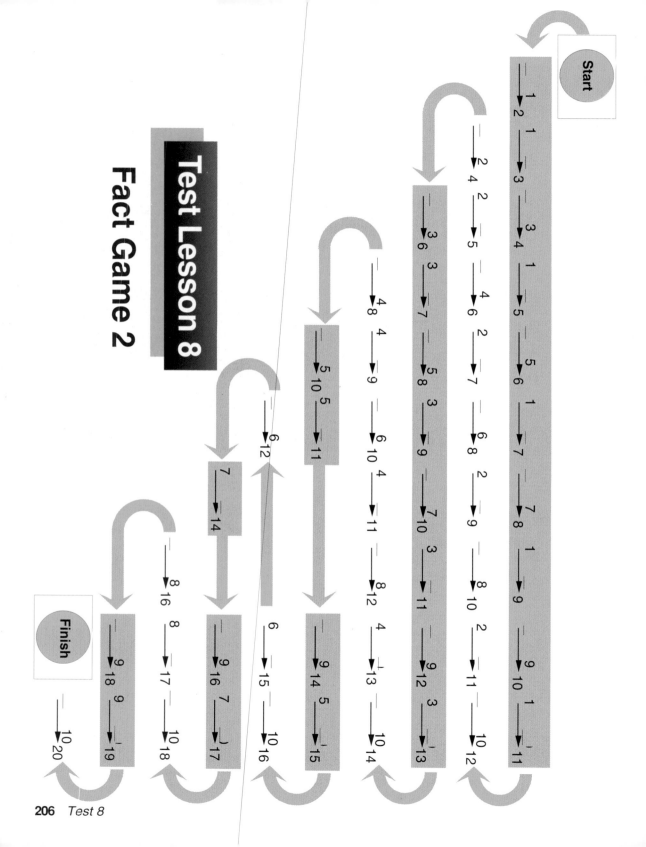

Fact Game 2

Test Lesson 8

Start

Finish

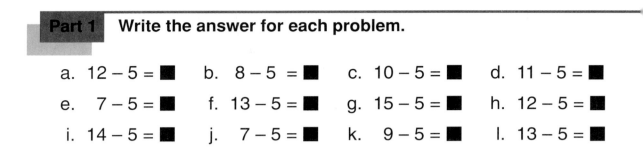

Lesson 81

Part 1 Write the answer for each problem.

a. $12 - 5 = $ ■   b. $8 - 5 = $ ■   c. $10 - 5 = $ ■   d. $11 - 5 = $ ■

e. $7 - 5 = $ ■   f. $13 - 5 = $ ■   g. $15 - 5 = $ ■   h. $12 - 5 = $ ■

i. $14 - 5 = $ ■   j. $7 - 5 = $ ■   k. $9 - 5 = $ ■   l. $13 - 5 = $ ■

 Part 2 Copy the table on lined paper.
Read each fact and put the numbers in the table.
Then figure out the missing numbers.

> The table is supposed to show the number of boys and girls that live on two streets.

	Boys	Girls	All children
Elm Street	37		
Oak Street			
Total for both streets			

Fact 1: 23 girls live on Elm Street.
Fact 2: 74 boys live on Oak Street.
Fact 3: The total number of girls is 42.
Fact 4: The total for all children is 153.

Part 3 Write the answer for each problem.

a. $5 \times 2 = $ ■   b. $9 \times 5 = $ ■   c. $4 \times 7 = $ ■   d. $3 \times 5 = $ ■

e. $5 \times 4 = $ ■   f. $4 \times 8 = $ ■   g. $9 \times 3 = $ ■   h. $9 \times 6 = $ ■

i. $3 \times 4 = $ ■   j. $9 \times 7 = $ ■   k. $5 \times 5 = $ ■   l. $9 \times 8 = $ ■

Write the estimation problem and the answer.

a. 7961	b. 7060	c. 9871	d. 8066	e. 9079
− 3056	− 5976	+ 2940	+ 1948	− 4980

Part 5

a. $\dfrac{2+5}{5}$ ---

b. $\dfrac{8-2}{7}$

c. $\dfrac{6+6}{10}$

d. $\dfrac{9-7}{5}$

e. $\dfrac{10-2}{8}$

f. $\dfrac{2+6}{9}$

Part 6

- You've learned a lot of subtraction facts. There's a way to check your answers to harder facts that start with a teens number.

- You look at the **difference** between the ones digits that are subtracted. Then you subtract that **difference from 10.**

- Here's a problem: $\begin{array}{r} 13 \\ -\ 5 \end{array}$

- You find the difference between the ones digits by subtracting. You start with the larger digit: **5 − 3 = 2**

- The difference is 2.

- You subtract that difference from 10: **10 − 2 = 8**

- The answer is 8.

- Here's another problem: $\begin{array}{r} 14 \\ -\ 8 \end{array}$

- The difference between 8 and 4 is 4: **8 − 4 = 4**

- Now you subtract from **10 − 4 = 6**. That's the answer.

a.   15
    − 8

b.   14
    − 9

c.   17
    − 9

Independent Work

Part 7   **Copy each problem and figure out the answer.**

a. 700
   x 9

b.   8
    x 4

c.   30
    x 4

d.   10
    x 7

e. 900
    x 4

Part 8   **Make a number family for each problem. Then write the addition or subtraction family and figure out the answer. Remember the unit name.**

a. Joe worked 273 fewer minutes than Sue worked. Sue worked 381 minutes. How many minutes did Joe work?

b. Andrea traveled 34 miles farther than Ginger traveled. Ginger traveled 120 miles. How many miles did Andrea travel?

c. Debbie started out with some bubbles. 85 bubbles popped. Debbie ended up with 223 bubbles. How many bubbles did Debbie start out with?

Write the answer for each problem.

a. 13
    − 3

b. 10
    − 5

c. 12
    − 9

d.   8
    − 5

e. 12
    − 3

f.   7
    − 1

g. 12
    − 8

h. 11
    − 8

i. 10
    − 7

j. 12
    − 6

k. 11
    − 3

l.   9
    − 9

Do the independent work for lesson 81 in your workbook.

Let's see how much cloth we need. 2 ft x 3 ft. How much is that?

Lesson 82

Part 1

- Sometimes not all the numbers are given in a table.
- There are facts that tell about other numbers. The facts tell you where to put the numbers in the table.

This table is supposed to show the number of small buildings and large buildings that were in two parks.

	Mountain Park	Valley Park	Total for both parks
Small buildings			
Large buildings	49		
All buildings			

Fact 1: There were 29 small buildings in Mountain Park.

Fact 2: There were 56 small buildings in Valley Park.

Fact 3: There were 80 large buildings in both parks.

Fact 4: There were 87 buildings in Valley Park.

a. Were there more small buildings or large buildings in Valley Park?

b. Were there more buildings in Mountain Park or in Valley Park?

c. How many buildings were there in both parks?

Part 2   Write the answer for each problem.

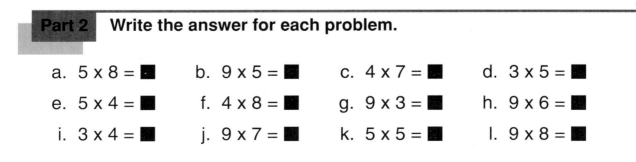

a. 5 x 8 = ■   b. 9 x 5 = ■   c. 4 x 7 = ■   d. 3 x 5 = ■

e. 5 x 4 = ■   f. 4 x 8 = ■   g. 9 x 3 = ■   h. 9 x 6 = ■

i. 3 x 4 = ■   j. 9 x 7 = ■   k. 5 x 5 = ■   l. 9 x 8 = ■

a. 6047	b. 696	c. 5968	d. 4006	e. 6982
+ 2001	+ 200	− 992	+ 2976	− 2973
8048	496	2077	6982	9955

Independent Work

Part 4 Use dollar amounts that are close to the values on the price tags to figure **about** how much. Use the values on the prices tags to figure **exactly** how much.

1	2	3	4	5
$2.94	$9.01	$1.03	$3.93	$3.04

a. A woman buys items 1, 3 and 5. **Exactly** how much does she spend?

b. A man buys items 2, 4 and 5. **About** how much does he spend?

c. A girl buys items 1, 2 and 4. **Exactly** how much does she spend?

Part 5 Make a number family for each problem. Then write the addition or subtraction problem for each family and figure out the answer. Remember the unit name.

a. A race car went 66 miles an hour faster than a train. The race car went 135 miles an hour. How fast did the train go?

b. A store had 465 golf balls. Then the store sold some golf balls. The store ended up with 128 golf balls. How many golf balls did the store sell?

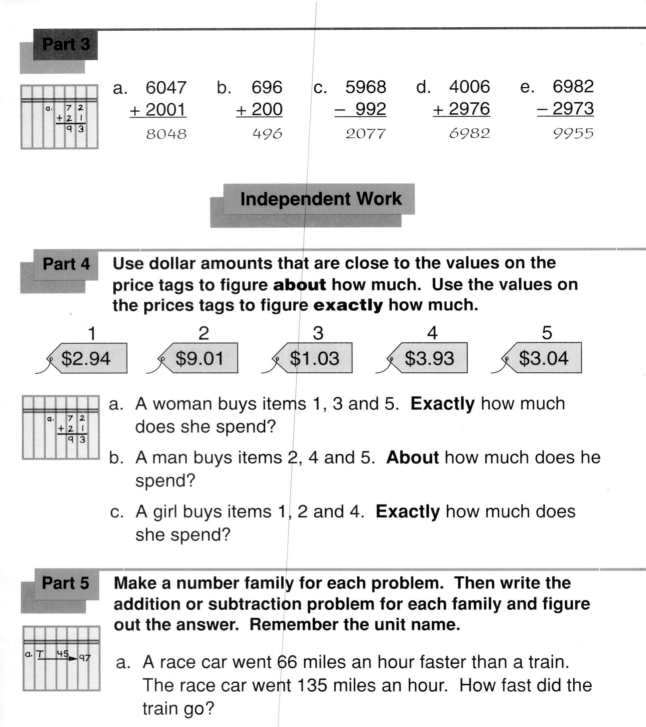

c. A sheep was 170 feet closer to the barn than a duck was. The duck was 199 feet from the barn. How far from the barn was the sheep?

d. A man had some money. Then he spend $6.50. He ended up with $5.19. How much money did he start out with?

Part 6 Write the time shown on each clock

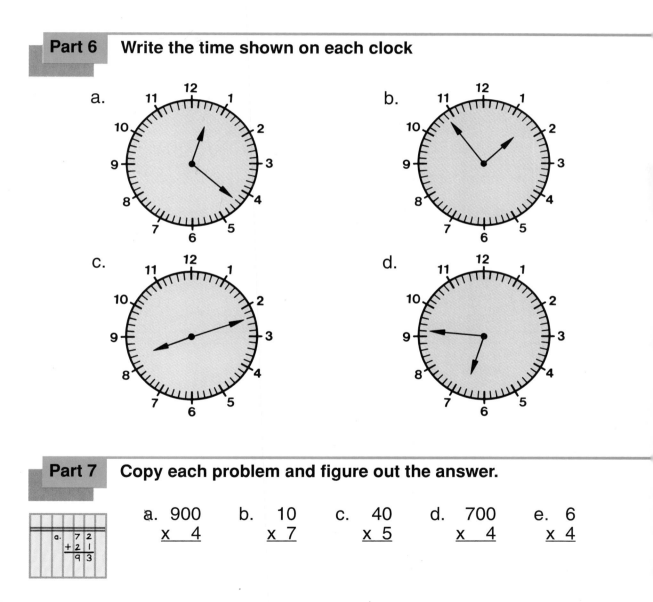

a.

b.

c.

d.

Part 7 Copy each problem and figure out the answer.

a. 900   b. 10   c. 40   d. 700   e. 6
  x   4   x 7   x 5   x   4   x 4

Lesson 83

Part 1 Write the answer to each problem.

a. 14 − 5 = ■   b. 13 − 5 = ■   c. 12 − 5 = ■   d. 12 − 7 = ■

e. 13 − 10 = ■   f. 13 − 9 = ■   g. 13 − 8 = ■   h. 15 − 9 = ■

i. 12 − 7 = ■   j. 12 − 6 = ■   k. 12 − 7 = ■   l. 13 − 8 = ■

Part 2

- Here's how to use estimation to find serious mistakes.
- Write the estimation problem and the answer for each problem the student worked.
- If the student's answer is way off, write the **exact** problem and the answer.
- Remember, don't work the exact problem if the student's answer is close to the estimation answer.

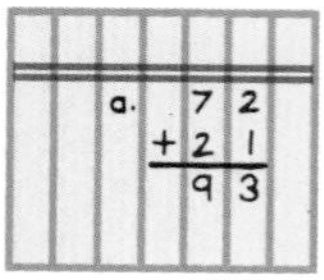

a.
```
 7 2
+2 1
 9 3
```

a. 9976   b. 3020   c. 9020   d. 4996
  − 4920   + 5940   − 7920   − 2003
   7056   8960   9940   2993

Copy the table on lined paper.
Read the facts and figure out all the numbers for the table.
Then answer the questions.

This table is supposed to show the number of green ducks and black ducks that were on two ponds.

	Mill Pond	Star Pond	Total for both ponds
Green ducks			
Black ducks		210	
Total ducks			

Fact 1: There was a total of 480 ducks on both ponds.

Fact 2: There was a total of 165 ducks on Mill Pond.

Fact 3: There was a total of 300 black ducks.

Fact 4: There were 75 green ducks on Mill Pond.

a. On Mill Pond, were there more black ducks or green ducks?
b. Which pond had more ducks?
c. Which pond had fewer green ducks?
d. How many green ducks were there on both ponds?

Independent Work

Part 4 **Copy each problem and figure out the answer.**

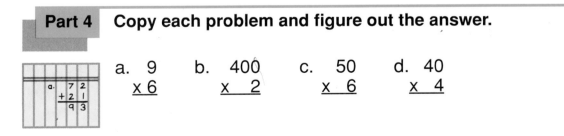

a. 9
   x 6

b. 400
   x   2

c. 50
   x 6

d. 40
   x 4

Write what X equals and what Y equals for each letter.

- Letter A. (X = ■, Y = ■)

- Letter B. (X = ■, Y = ■)

- Letter C. (X = ■, Y = ■)

- Letter D. (X = ■, Y = ■)

Lesson 84

Write the answer for each problem.

a. $6 \times 9 = $ 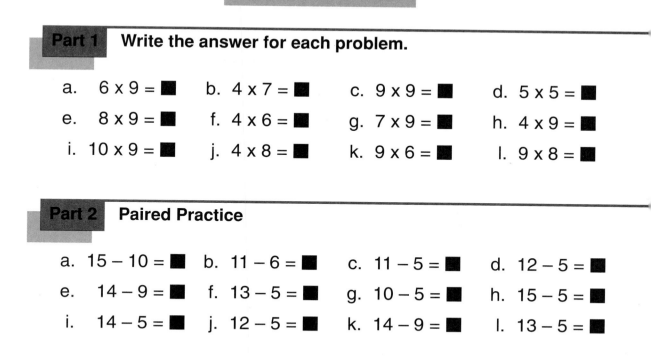   b. $4 \times 7 = $ c. $9 \times 9 = $ d. $5 \times 5 = $

e. $8 \times 9 = $ f. $4 \times 6 = $ g. $7 \times 9 = $ h. $4 \times 9 = $

i. $10 \times 9 = $ j. $4 \times 8 = $ k. $9 \times 6 = $ l. $9 \times 8 = $

Part 2 **Paired Practice**

a. $15 - 10 = $ b. $11 - 6 = $ c. $11 - 5 = $ d. $12 - 5 = $

e. $14 - 9 = $ f. $13 - 5 = $ g. $10 - 5 = $ h. $15 - 5 = $

i. $14 - 5 = $ j. $12 - 5 = $ k. $14 - 9 = $ l. $13 - 5 = $

Part 3

- Here's 47 times 5. When you work the multiplication problem, you actually work two multiplication problems.

- The arrows show that there's a problem for the ones and a problem for the tens.

$$\begin{array}{r} 47 \\ \times\ 5 \\ \hline \end{array}$$

$$\begin{array}{cc} 7 & 40 \\ \times 5 & \times 5 \\ \hline 35 & + \quad 200 \end{array} = 235$$

Part 4 Copy the table on lined paper.
Read the facts and figure out all the numbers for the table.
Then answer the questions.

This table is supposed to show the number of trucks and cars that traveled on Oak Street and Maple Lane.

	Trucks	Cars	Total Vehicles
Oak Street			95
Maple Lane			
Total for both streets			

Fact 1: There were 382 vehicles that traveled on both streets.

Fact 2: There were 211 cars that traveled on Maple Lane.

Fact 3: There were 76 trucks that traveled on Maple Lane.

Fact 4: There were 95 cars that traveled on Oak Street.

a. How many trucks travelled on Oak Street?
b. Did more vehicles travel on Oak Street or Maple Lane?
c. Were there more trucks on both streets or more cars on both streets?

Part 5

Use dollar amounts that are close to the values on the price tags to figure **about** how much. Use the values on the price tags to figure **exactly** how much.

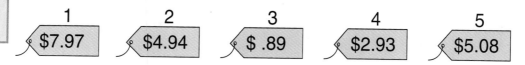

1	2	3	4	5
$7.97	$4.94	$ .89	$2.93	$5.08

a. The girl buys items 1 and 5. **Exactly** how much does she spend?

b. A man buys items 1, 3 and 4. **Exactly** how much does he spend?

c. A woman buys items 2, 4 and 5. **About** how much does she spend?

d. A boy buys items 1, 2 and 4. **About** how much does he spend?

Part 6

Make a number family for each problem. Then write the addition or subtraction problem and figure out the answer. Remember the unit name.

a. A cow had some oats. The cow ate 106 pounds of oats. The cow ended up with 181 pounds of oats. How many pounds of oats did the cow start with?

b. Jan had 76 fewer stamps than Donna. Donna has 330 stamps. How many stamps did Jan have?

c. A rat collected 78 fewer nuts than a bird collected. The bird collected 323 nuts. How many nuts did the rat collect?

d. Fran had 498 stamps. She sold some stamps. She ended up with 67 stamps. How many stamps did she sell?

Lesson 85

Write the answer for each problem.

a. $15 - 6 = \blacksquare$ b. $10 - 6 = \blacksquare$ c. $13 - 6 = \blacksquare$ d. $14 - 6 = \blacksquare$

e. $12 - 6 = \blacksquare$ f. $6 - 6 = \blacksquare$ g. $16 - 6 = \blacksquare$ h. $9 - 6 = \blacksquare$

Part 2

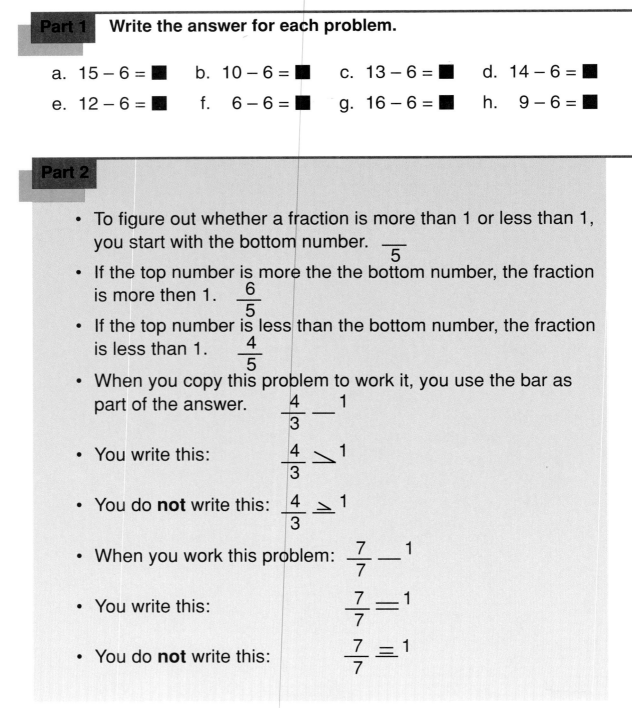

- To figure out whether a fraction is more than 1 or less than 1, you start with the bottom number. $\dfrac{}{5}$
- If the top number is more the the bottom number, the fraction is more then 1. $\dfrac{6}{5}$
- If the top number is less than the bottom number, the fraction is less than 1. $\dfrac{4}{5}$
- When you copy this problem to work it, you use the bar as part of the answer. $\dfrac{4}{3}$ __ 1
- You write this: $\dfrac{4}{3} > 1$
- You do **not** write this: $\dfrac{4}{3} \geq 1$
- When you work this problem: $\dfrac{7}{7}$ __ 1
- You write this: $\dfrac{7}{7} = 1$
- You do **not** write this: $\dfrac{7}{7} = 1$

a. $\dfrac{7}{6}$ __ 1

b. $\dfrac{8}{8}$ __ 1

c. 1 __ $\dfrac{3}{4}$

d. $\dfrac{5}{5}$ __ 1

e. $\dfrac{7}{8}$ __ 1

f. $\dfrac{4}{3}$ __ 1

Independent Work

Part 3 Use dollar amounts that are close to the values on the price tags to figure **about** how much. Use the values on the price tags to figure **exactly** how much.

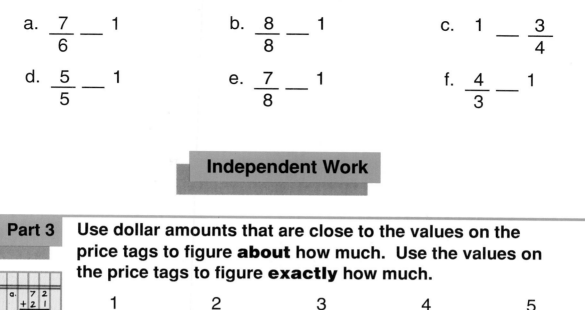

a. $\dfrac{7\ 2}{+2\ 1}$
  $\overline{9\ 3}$

1 $3.58
2 $6.03
3 $ .95
4 $4.94
5 $5.75

a. A man buys items 2, 3 and 4. **About** how much does he spend?

b. A girl buys items 1, 3 and 5. **Exactly** how much does she spend?

c. A woman buys items 2, 3 and 5. **Exactly** how much does she spend?

d. A boy buys items 2 and 3. **About** how much does he spend?

- Each problem shows some coins and a number family with letters.

- Each dime is worth 10 cents. That's the first number in the family.

- The other small number in the family is D. That's the number of dimes. There are 3 dimes, so you cross out D and write 3.

- Now you figure out the big number. That's the number of cents you have for 3 dimes— 10 x 3 = 30. So you cross out C and write 30.

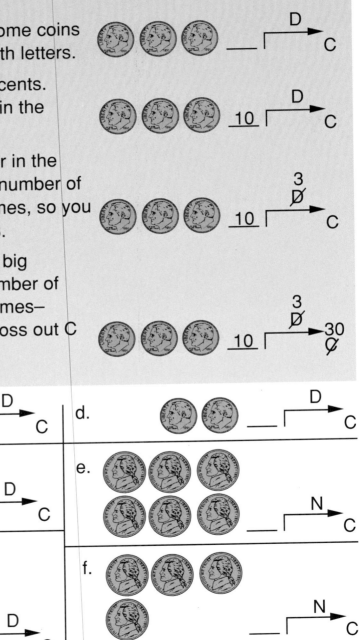

a.

b.

c.

d.

e.

f.

Part 2 Write answers to all the problems.

   a. 15 − 6 = ■   b. 14 − 6 = ■   c. 13 − 6 = ■   d. 12 − 6 = ■

   e.   9 − 6 = ■   f. 16 − 6 = ■   g. 13 − 6 = ■   h. 15 − 6 = ■

   i. 14 − 6 = ■

Independent Work

Part 3 Copy each problem. Complete the sign.

   a. $\dfrac{8}{3}$ __ $\dfrac{7}{3}$ b. $\dfrac{9}{6}$ __ $\dfrac{5}{6}$ c. $\dfrac{3}{4}$ __ $\dfrac{4}{4}$

Part 4 Make a number family for each problem. Then write the addition or subtraction problem and figure out the answer. Remember the unit name.

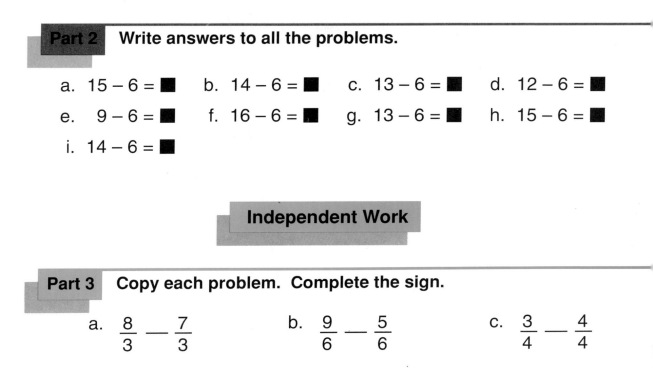

a. Jan had some seeds. She planted 49 seeds. She ended up with 236 seeds. How many seeds did she start out with?

b. A girl had some toys. She gave away 92 toys. She ended up with 211 toys. How many toys did she have at first?

c. The pole was 35 feet shorter than the tree. The pole was 97 feet tall. How tall was the tree?

d. George weighed 330 pounds less than a horse. The horse weighed 520 pounds. How many pounds did George weigh?

Part 5 Write the fraction for each problem.

a.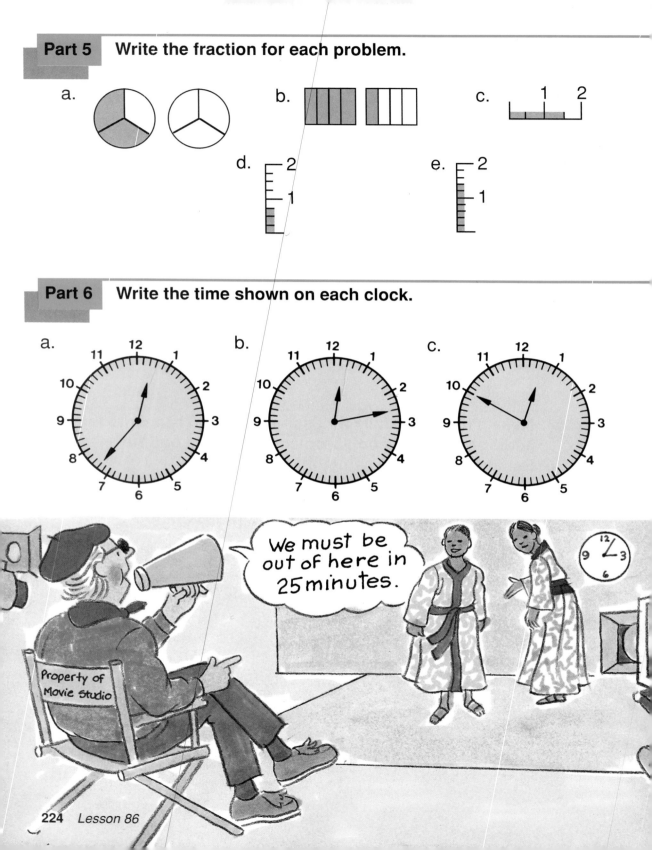

b.

c.

d.

e.

Part 6 Write the time shown on each clock.

a.

b.

c.

We must be out of here in 25 minutes.

Property of Movie Studio

Part 1

- The first number you put in the family tells how many cents each coin is worth.

- If the coins are dimes, the first small number is 10. Then you count the dimes and replace D with that number.

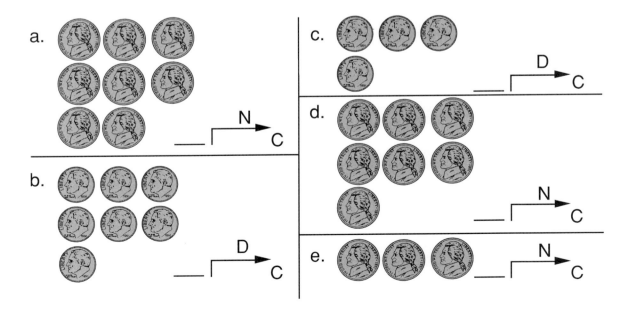

a.

```
    ___  |⌐  N
           └─── C
```

b.

```
    ___  |⌐  D
           └─── C
```

c.

```
    ___  |⌐  D
           └─── C
```

d.

```
    ___  |⌐  N
           └─── C
```

e.

```
    ___  |⌐  N
           └─── C
```

Write the answer to each problem.

a. $12 - 6 =$ ■   b. $10 - 6 =$ ■   c. $14 - 6 =$ ■   d. $16 - 6 =$ ■

e. $13 - 6 =$ ■   f. $11 - 6 =$ ■   g. $9 - 6 =$ ■   h. $14 - 6 =$ ■

i. $13 - 6 =$ ■

Independent Work

Part 3

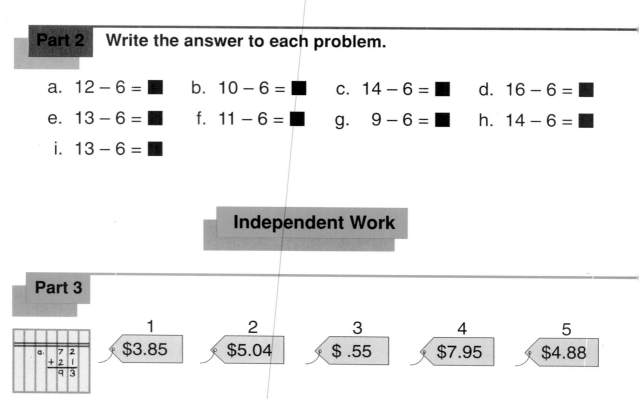

1	2	3	4	5
$3.85	$5.04	$ .55	$7.95	$4.88

a. The girl buys items 1, 2 and 3. **Exactly** how much does she spend?

b. A man buys items 2, 4 and 5. **About** how much does he spend?

c. The girl buys item 4. **About** how much does she spend?

d. The woman buys items 1, 2 and 5. **About** how much does she spend?

e. The boy buys items 3, 4 and 5. **Exactly** how much does he spend?

Part 4   Write what X equals and what Y equals for each letter.

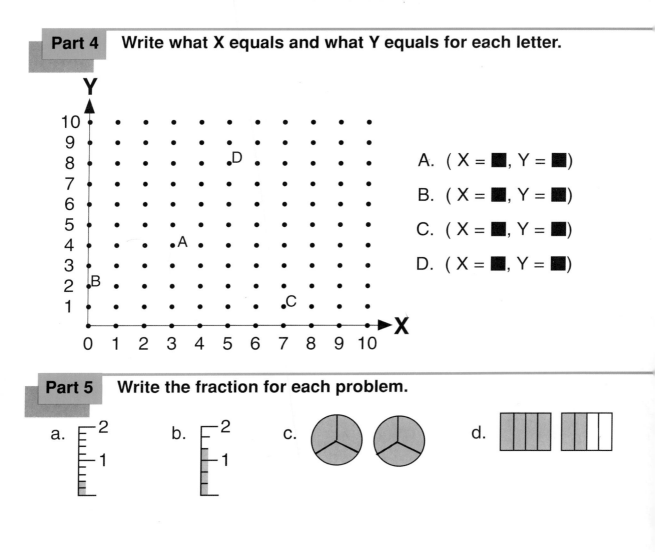

A. (X = ■, Y = ■)

B. (X = ■, Y = ■)

C. (X = ■, Y = ■)

D. (X = ■, Y = ■)

Part 5   Write the fraction for each problem.

a.   2   1

b.   2   1

c.

d.

Lesson 88

Part 1

• For column multiplication problems, first you work the problem in the ones column.

$$\begin{array}{r} 36 \\ \times\ 4 \\ \hline \end{array}$$

• If the answer has more than one digit you carry the first digit.

$$\begin{array}{r} \overset{2}{3}6 \\ \times\ 4 \\ \hline 4 \end{array}$$

• Then you work the tens. You multiply. Then you add the number you carried.

$$\begin{array}{r} \overset{2}{3}6 \\ \times\ 4 \\ \hline 144 \end{array}$$

a.   42
   x 6

b.   27
   x 4

c.   76
   x 9

d.   95
   x 2

e.   32
   x 9

Write the answer to each problem.

a. $\begin{array}{r} 14 \\ -\ 6 \\ \hline \end{array}$ b. $\begin{array}{r} 12 \\ -\ 6 \\ \hline \end{array}$ c. $\begin{array}{r} 13 \\ -\ 6 \\ \hline \end{array}$ d. $\begin{array}{r} 15 \\ -\ 6 \\ \hline \end{array}$ e. $\begin{array}{r} 11 \\ -\ 6 \\ \hline \end{array}$

f. $\begin{array}{r} 13 \\ -\ 7 \\ \hline \end{array}$ g. $\begin{array}{r} 16 \\ -10 \\ \hline \end{array}$ h. $\begin{array}{r} 14 \\ -\ 8 \\ \hline \end{array}$ i. $\begin{array}{r} 15 \\ -\ 9 \\ \hline \end{array}$ j. $\begin{array}{r} 16 \\ -\ 6 \\ \hline \end{array}$

Independent Work

Part 3 Use dollar amounts that are close to the values on the price tags to figure **about** how much. Use the values on the prices tags to figure **exactly** how much.

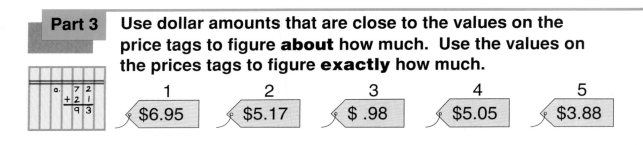

1 $6.95   2 $5.17   3 $ .98   4 $5.05   5 $3.88

a. A boy buys items 1, 2 and 3. **Exactly** how much does he spend?

b. A girl buys items 1, 4 and 5. **Exactly** how much does she spend?

c. The woman buys items 2 and 4. **About** how much does she spend?

d. A man buys items 2, 4 and 5. **Exactly** how much does he spend?

Part 4 Copy the problems you can work and work them.

a. $\dfrac{7}{3} - \dfrac{5}{2} =$

b. $\dfrac{2}{6} + \dfrac{5}{6} =$

c. $\dfrac{8}{2} - \dfrac{3}{2} =$

d. $\dfrac{11}{4} + \dfrac{3}{4} =$

e. $\dfrac{8}{10} - \dfrac{3}{11} =$

f. $\dfrac{12}{9} - \dfrac{5}{9} =$

Part 5

Make a number family for each problem. Then write the addition or subtraction problem for each family and figure out the answer. Remember the unit name.

a. A man ran 185 feet farther than his dog. The man ran 932 feet. How far did the dog run?

b. A maple tree was 39 feet shorter than an oak tree. The oak tree was 114 feet tall. How tall was the maple tree?

c. A rope was 118 feet shorter than a chain. The chain was 167 feet long. How long was the rope?

d. Ginger had lots of stamps. Then she bought 31 more stamps. She ended up with 373 stamps. How many stamps did she start out with?

Part 6

Write the fraction for each problem.

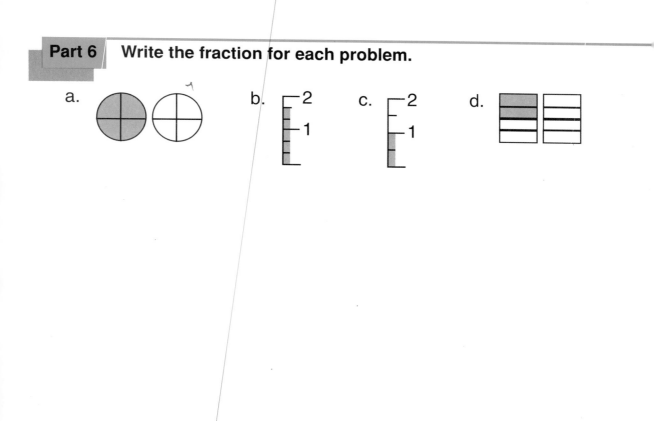

a.

b. 2 1

c. 2 1

d.

Lesson 89

Part 1

- Here's how you add and subtract fractions that have the same bottom number: First write the bottom number. Then figure out the top number.

a. $\dfrac{17}{3} - \dfrac{9}{3} =$

b. $\dfrac{13}{10} - \dfrac{4}{10} =$

c. $\dfrac{12}{7} + \dfrac{3}{7} =$

d. $\dfrac{2}{7} + \dfrac{14}{7} =$

e. $\dfrac{3}{8} + \dfrac{10}{8} =$

f. $\dfrac{18}{2} - \dfrac{9}{2} =$

Part 2 Write the answer to each problem.

a. $\begin{array}{r} 14 \\ -\ 6 \\ \hline \end{array}$

b. $\begin{array}{r} 15 \\ -\ 5 \\ \hline \end{array}$

c. $\begin{array}{r} 14 \\ -\ 9 \\ \hline \end{array}$

d. $\begin{array}{r} 13 \\ -\ 7 \\ \hline \end{array}$

e. $\begin{array}{r} 13 \\ -\ 9 \\ \hline \end{array}$

f. $\begin{array}{r} 13 \\ -\ 4 \\ \hline \end{array}$

g. $\begin{array}{r} 13 \\ -\ 6 \\ \hline \end{array}$

h. $\begin{array}{r} 16 \\ -\ 9 \\ \hline \end{array}$

i. $\begin{array}{r} 16 \\ -10 \\ \hline \end{array}$

j. $\begin{array}{r} 14 \\ -\ 8 \\ \hline \end{array}$

k. $\begin{array}{r} 17 \\ -\ 9 \\ \hline \end{array}$

l. $\begin{array}{r} 16 \\ -\ 7 \\ \hline \end{array}$

Part 3

a. $\begin{array}{r} 49 \\ \times\ 3 \\ \hline \end{array}$

b. $\begin{array}{r} 92 \\ \times\ 6 \\ \hline \end{array}$

c. $\begin{array}{r} 37 \\ \times\ 4 \\ \hline \end{array}$

d. $\begin{array}{r} 22 \\ \times\ 9 \\ \hline \end{array}$

e. $\begin{array}{r} 54 \\ \times\ 5 \\ \hline \end{array}$

Part 4 Make a number family for each problem.
Then write the addition or subtraction problem for each
family and figure out the answer.

a. Joe has $4.50 less than Steve. Steve has $9.32. How
much money does Joe have?

b. Joe had some money. He spent $5.83. He ended up
with $6.00. How much money did he start with?

Part 5 Write the fraction for each problem.

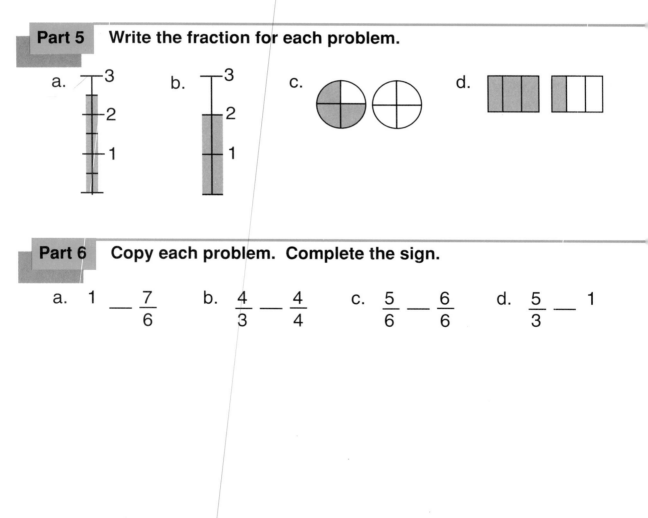

a.

```
─ 3
─ 2
─ 1
```

b.

```
─ 3
─ 2
─ 1
```

c.

d.

Part 6 Copy each problem. Complete the sign.

a. $1 \underline{} \dfrac{7}{6}$ b. $\dfrac{4}{3} \underline{} \dfrac{4}{4}$ c. $\dfrac{5}{6} \underline{} \dfrac{6}{6}$ d. $\dfrac{5}{3} \underline{} 1$

Lesson 90

Part 1

- These are word problems that have fractions. They work just like other problems.

- Here's a problem: Carol ran $\frac{3}{5}$ of a mile. Then she ran $\frac{4}{5}$ of a mile. How far did she run in all?

$$\frac{3}{5} + \frac{4}{5} = \frac{7}{5}$$

- The fractions are shown for this problem. 3-fifths goes first. Then plus 4-fifths. 7-fifths is how many miles she ran in all.

a. Carol ran $\frac{2}{8}$ of a mile. Then she ran $\frac{3}{8}$ of a mile. How many miles did she run in all?

b. The boys ate $\frac{1}{7}$ of a pie. Then they ate $\frac{6}{7}$ of a pie. How many pies did they eat in all?

c. The girls ate $\frac{3}{7}$ of a pie. Then they ate $\frac{4}{7}$ of a pie. How many pies did they eat in all?

Part 2

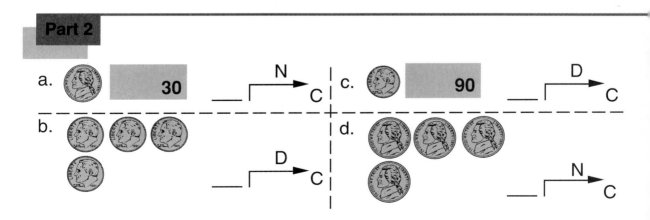

Copy the problems and work them.

a.   12   b.   96   c.   52   d.   34   e.   27   f.   34
    x 5   x 2   x 9   x 5   x 4   x 4

Part 4   **Paired Practice**

a.   15   b.   13   c.   14   d.   13   e.   13   f.   14
    – 9   – 7   – 9   – 6   – 9   – 6

g.   12   h.   15   i.   16   j.   14   k.   13   l.   13
    – 6   – 6   – 6   – 8   – 6   – 7

Independent Work

Part 5   **Use dollar amounts that are close to the values on the price tags to figure about how much. Use the values on the prices tags to figure exactly how much.**

1	2	3	4	5
$ .96	$4.88	$4.07	$2.92	$8.11

a. A man buys items 1, 2 and 4. **Exactly** how much does he spend?

b. The woman buys items 3 and 5. **About** how much does she spend?

c. A girl buys items 2, 3 and 5. **About** how much does she spend?

d. The woman buys items 2, 3 and 5. **Exactly** how much does she spend?

Part 6　Copy each problem and work it.

a. $\dfrac{12}{5} - \dfrac{5}{5} =$

b. $\dfrac{4}{8} + \dfrac{3}{8} =$

c. $\dfrac{9}{2} - \dfrac{3}{2} =$

d. $\dfrac{2}{3} + \dfrac{15}{3} =$

Part 7　Copy the problems you can work. Then work those problems.

a. $\dfrac{8}{13} + \dfrac{8}{8} =$

b. $\dfrac{6}{3} - \dfrac{5}{3} =$

c. $\dfrac{12}{5} - \dfrac{5}{5} =$

d. $\dfrac{9}{2} - \dfrac{2}{3} =$

e. $\dfrac{6}{4} + \dfrac{5}{6} =$

f. $\dfrac{2}{3} + \dfrac{15}{3} =$

Test 9

Part 1

a. $7 \times 9 = \blacksquare$ b. $4 \times 9 = \blacksquare$ c. $9 \times 9 = \blacksquare$ d. $5 \times 5 = \blacksquare$

e. $8 \times 9 = \blacksquare$ f. $4 \times 8 = \blacksquare$ g. $6 \times 9 = \blacksquare$ h. $4 \times 7 = \blacksquare$

i. $10 \times 9 = \blacksquare$ j. $4 \times 6 = \blacksquare$ k. $9 \times 8 = \blacksquare$ l. $9 \times 6 = \blacksquare$

Part 2

a. $14 - 6$ b. $12 - 6$ c. $13 - 6$ d. $15 - 6$ e. $11 - 6$

f. $13 - 7$ g. $16 - 10$ h. $14 - 8$ i. $15 - 9$ j. $16 - 6$

Part 3 Write the estimation problem and the answer.

a. $9976 - 4920$ b. $3020 + 8940$ c. $16020 - 7980$ d. $12996 - 8072$

Part 4 Copy each problem and work it.

a. 49×3 b. 92×6 c. 37×4 d. 22×9

Part 5 Copy the problems you can work. Then work those problems.

a. $\dfrac{3}{4} + \dfrac{2}{3}$ b. $\dfrac{3}{10} + \dfrac{9}{10}$ c. $\dfrac{17}{5} - \dfrac{8}{9}$ d. $\dfrac{10}{3} - \dfrac{9}{3}$

Go to Test 9 in your workbook.

Test Lesson 9

Fact Game 1

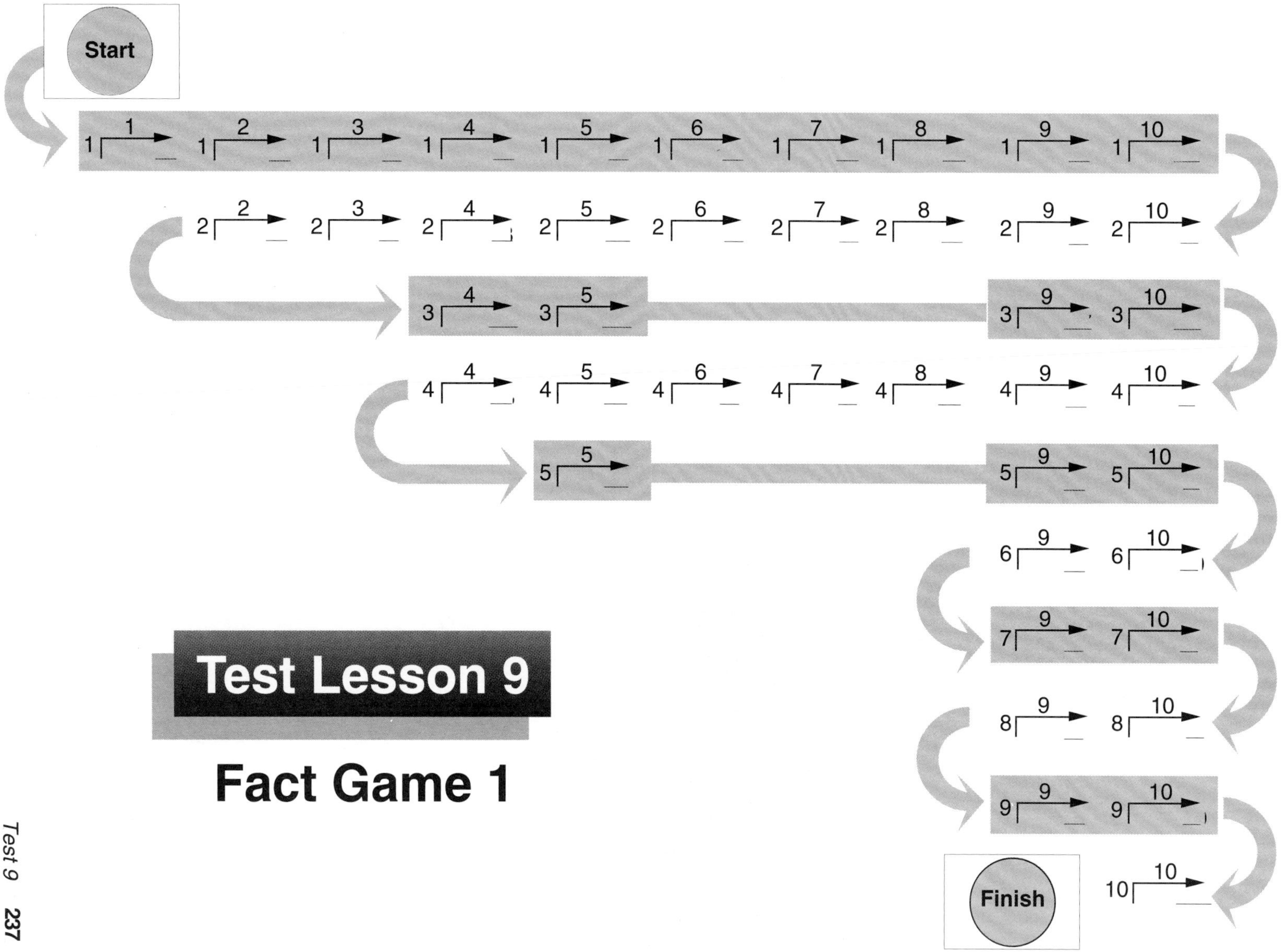

Start

Finish

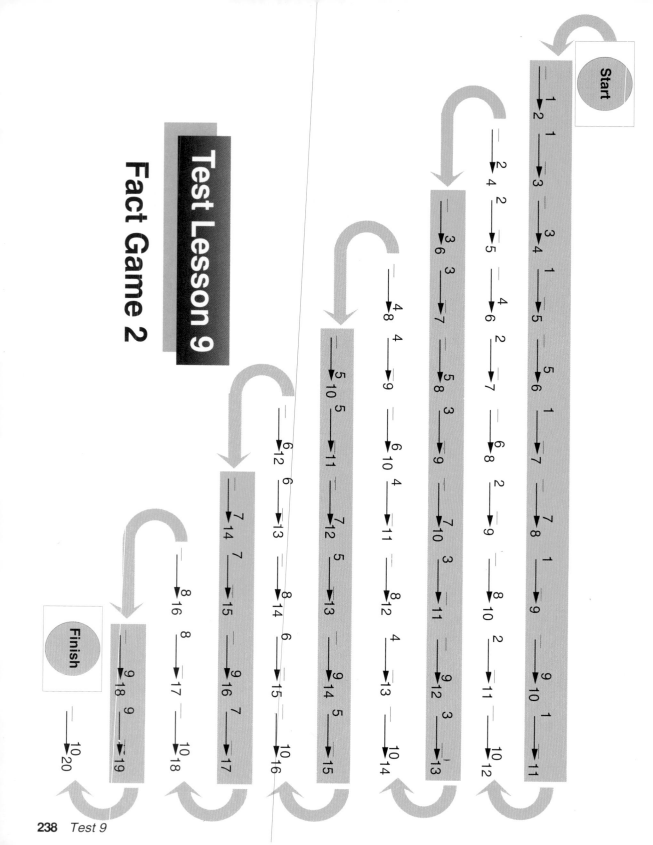

Fact Game 2

Test Lesson 9

Start

Finish

Lesson 91

Part 1 Write the answers to the problems.

a. $5 \times 6 = $ ■   b. $5 \times 8 = $ ■   c. $5 \times 7 = $ ■   d. $5 \times 9 = $ ■

e. $5 \times 5 = $ ■   f. $5 \times 3 = $ ■   g. $5 \times 6 = $ ■   h. $5 \times 10 = $ ■

i. $5 \times 8 = $ ■   j. $5 \times 7 = $ ■

Part 2 Write the number problems and the answers. Remember the unit name.

a. A rabbit weighed $\frac{8}{3}$ pounds. Then the rabbit lost $\frac{2}{3}$ of a pound. How many pounds did the rabbit end up weighing?

b. A bag of nails weighed $\frac{9}{4}$ pounds. Somebody took $\frac{5}{4}$ pounds of nails from the bag. How many pounds were left in the bag?

c. A rope was $\frac{6}{4}$ feet long. Somebody cut $\frac{1}{4}$ foot from the rope. How many feet of rope were left?

d. A rope was $\frac{4}{5}$ foot long. Somebody made the rope $\frac{3}{5}$ foot longer. How long was the rope?

Part 3 Write the answers to the problems.

a. $15 - 7 = $ ■   b. $15 - 8 = $ ■   c. $15 - 9 = $ ■   d. $15 - 10 = $ ■

e. $14 - 7 = $ ■   f. $14 - 8 = $ ■   g. $14 - 9 = $ ■   h. $14 - 10 = $ ■

i. $16 - 9 = $ ■   j. $16 - 10 = $ ■   k. $15 - 8 = $ ■

- Some of these problems show the number of coins. Some show how many cents the coins are worth.

- This problem shows how many cents the nickels are worth:

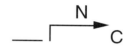 **20**

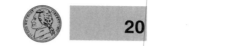

So you write 5 as the first small number because each nickel is worth 5 cents. You write 20 for cents. Then you can work the problem to find the number of nickels.

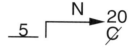

- This problem shows how many nickels are in the group:

So you write 5 as the first small number because each nickel is worth 5 cents. You write 3 for the number of nickels. Then you can work the problem to figure out the number of cents.

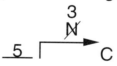

Part 5 Copy each number family. Put in the numbers and figure out the answers.

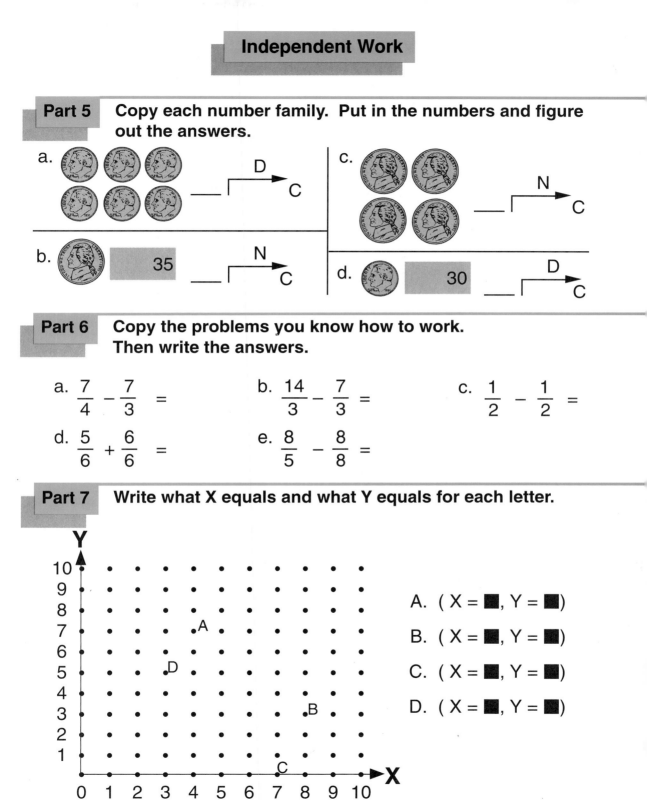

a. _____ ⌐D→ C

b. 35 _____ ⌐N→ C

c. _____ ⌐N→ C

d. 30 _____ ⌐D→ C

Part 6 Copy the problems you know how to work.
Then write the answers.

a. $\dfrac{7}{4} - \dfrac{7}{3} =$

b. $\dfrac{14}{3} - \dfrac{7}{3} =$

c. $\dfrac{1}{2} - \dfrac{1}{2} =$

d. $\dfrac{5}{6} + \dfrac{6}{6} =$

e. $\dfrac{8}{5} - \dfrac{8}{8} =$

Part 7 Write what X equals and what Y equals for each letter.

A. (X = ■, Y = ■)

B. (X = ■, Y = ■)

C. (X = ■, Y = ■)

D. (X = ■, Y = ■)

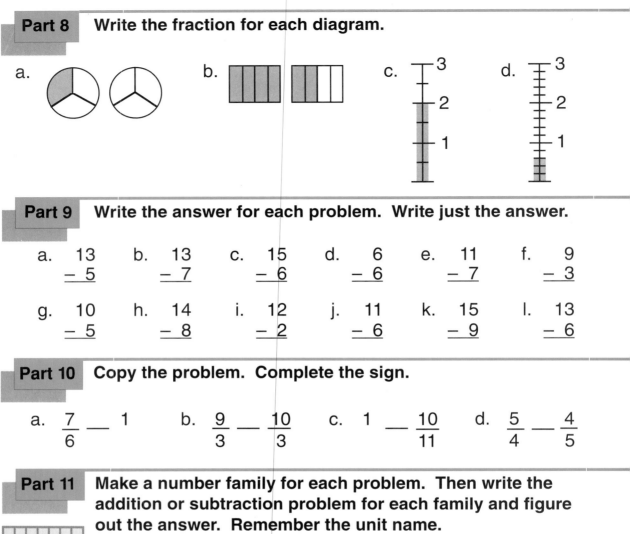

Part 8 Write the fraction for each diagram.

a.

b.

c.

d.

Part 9 Write the answer for each problem. Write just the answer.

a. 13
  − 5

b. 13
  − 7

c. 15
  − 6

d. 6
  − 6

e. 11
  − 7

f. 9
  − 3

g. 10
  − 5

h. 14
  − 8

i. 12
  − 2

j. 11
  − 6

k. 15
  − 9

l. 13
  − 6

Part 10 Copy the problem. Complete the sign.

a. $\dfrac{7}{6}$ __ 1

b. $\dfrac{9}{3}$ __ $\dfrac{10}{3}$

c. 1 __ $\dfrac{10}{11}$

d. $\dfrac{5}{4}$ __ $\dfrac{4}{5}$

Part 11 Make a number family for each problem. Then write the addition or subtraction problem for each family and figure out the answer. Remember the unit name.

a. In the morning, Jan drove some miles. In the afternoon, she drove 126 more miles. She ended up 213 miles from home. How many miles did she drive in the morning?

b. A grocery store had lots of eggs. The store sold 345 eggs. The store ended up with 661 eggs. How many eggs did the store start out with?

c. A horse weighed 3200 pounds less than an elephant. The horse weighed 1560 pounds. How many pounds did the elephant weigh?

Lesson 92

Part 1 Copy each problem and work it.

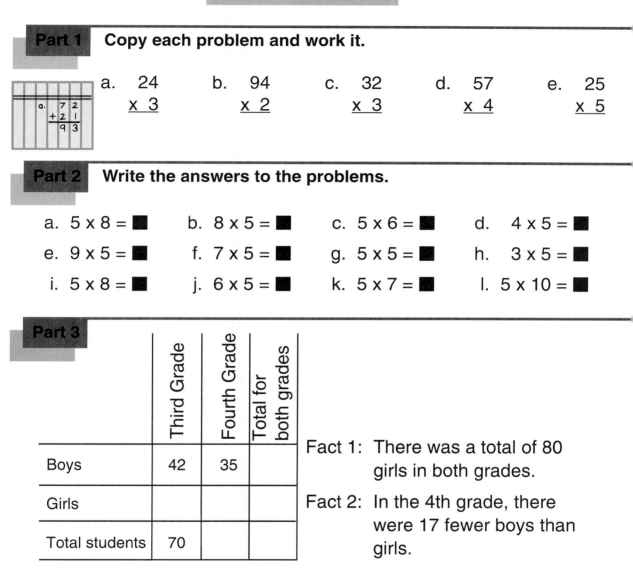

a.	24	b.	94	c.	32	d.	57	e.	25
	x 3		x 2		x 3		x 4		x 5

Part 2 Write the answers to the problems.

a. 5 x 8 = ■   b. 8 x 5 = ■   c. 5 x 6 = ■   d. 4 x 5 = ■

e. 9 x 5 = ■   f. 7 x 5 = ■   g. 5 x 5 = ■   h. 3 x 5 = ■

i. 5 x 8 = ■   j. 6 x 5 = ■   k. 5 x 7 = ■   l. 5 x 10 = ■

Part 3

	Third Grade	Fourth Grade	Total for both grades
Boys	42	35	
Girls			
Total students	70		

Fact 1: There was a total of 80 girls in both grades.

Fact 2: In the 4th grade, there were 17 fewer boys than girls.

a. How many boys were in both grades?

b. How many girls were in the third grade?

c. Were there more students in the third grade or the fourth grade?

d. Were there more boys in the third grade or in the fourth grade?

e. How many total students were in both grades?

Part 4   **Write the answers to the problems.**

a. $15 - 8 = \blacksquare$ b. $15 - 9 = \blacksquare$ c. $13 - 7 = \blacksquare$ d. $13 - 6 = \blacksquare$

e. $13 - 8 = \blacksquare$ f. $13 - 9 = \blacksquare$ g. $13 - 10 = \blacksquare$ h. $15 - 7 = \blacksquare$

i. $15 - 8 = \blacksquare$ j. $14 - 7 = \blacksquare$ k. $16 - 7 = \blacksquare$

Part 5

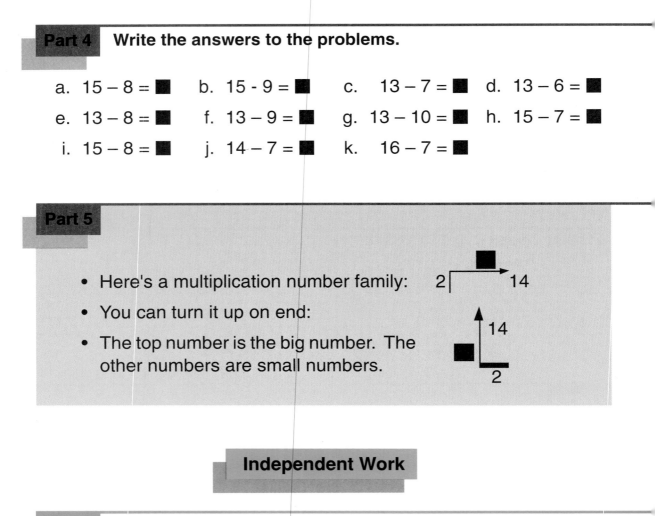

- Here's a multiplication number family:
- You can turn it up on end:
- The top number is the big number. The other numbers are small numbers.

Independent Work

Part 6   **Write the number problems and the answers. Remember the unit name.**

a. A baker had $\frac{9}{4}$ cups of sugar. She used $\frac{3}{4}$ cup of sugar. How much sugar did she end up with?

b. A chicken weighed $\frac{9}{2}$ pounds. The chicken gained $\frac{1}{2}$ pound. How much did the chicken end up weighing?

c. A baby started with $\frac{8}{8}$ of a bottle. The baby drank $\frac{3}{8}$ of the bottle. How much of the bottle did the baby have left?

d. Joe had $\frac{7}{3}$ bags of peanuts. He bought $\frac{6}{3}$ bags of peanuts. How many bags of peanuts did Joe end up with?

Part 7 Write the fraction for each diagram.

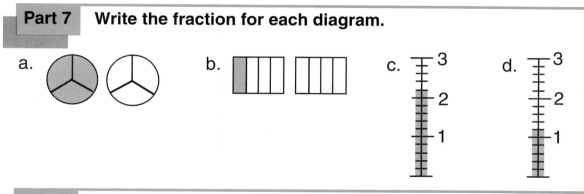

a. b. c. $\frac{}{}$ 3 d. $\frac{}{}$ 3

Part 8 Make a number family for each problem. Then write the number problem and the answer. Remember the unit name.

a. Steve is 473 millimeters shorter than Dave. Dave is 1931 millimeters tall. How tall is Steve?

b. There were 126 fewer people in the park than there was at the beach. 781 people were at the beach. How many people were in the park?

c. Jane had lots of paper. She used up 111 sheets of paper. She ended up with 12 sheets of paper. How many sheets did she start out with?

d. Mr. Wilson had 1477 nails. Then he bought some more nails. He ended up with 3820 nails. How many nails did he buy?

Part 9 Copy and work the problems you know how to work.

a. $\frac{14}{2} + \frac{1}{2} =$ b. $\frac{13}{13} - \frac{13}{5} =$ c. $\frac{3}{12} + \frac{10}{12} =$ d. $\frac{4}{8} + \frac{8}{8} =$

Part 10 Copy each problem and work it.

a.   144
   − 28

b.   637
   103
   + 26

c.   503
   − 455

d.   738
   + 68

Part 1

- Here's a problem: **There were 60 cars in all. 14 were red cars. The rest were blue cars. How many blue cars were there?**

- You write **R** for **red** cars, **B** for **blue** cars and **All** for all cars. The word **cars** goes under the number family.

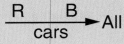

$$\overset{\text{R} \qquad \text{B}}{\underset{\text{cars}}{\rule{3cm}{0.4pt}}}\!\!\!\longrightarrow \text{All}$$

- Then you put in the numbers you know: 14 were red cars. There were 60 cars in all.

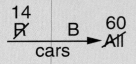

$$\overset{\overset{14}{\text{R̶}} \qquad \text{B} \quad \overset{60}{}}{\underset{\text{cars}}{\rule{3cm}{0.4pt}}}\!\!\!\longrightarrow \text{A̶l̶l̶}$$

- Now you have a number family with two numbers. You can figure out the missing number.

a. Tim had **red** marbles and **green** marbles. He had 14 red marbles. He had 16 green marbles. How many marbles did he have in all?

b. Dora had **big** jars and **little** jars. She had 35 jars in all. 14 were big jars. How many were little jars?

Part 2   **Write the answers to the problems.**

a. 10 x 5 = ■   b. 8 x 5 = ■   c. 5 x 5 = ■   d. 7 x 5 = ■

e. 5 x 4 = ■   f. 5 x 2 = ■   g. 5 x 8 = ■   h. 9 x 5 = ■

i. 5 x 7 = ■   j. 5 x 6 = ■   k. 5 x 8 = ■   l. 6 x 5 = ■

Copy each problem and work it.

a. 678
   x 2

b. 435
   x 5

c. 938
   x 4

d. 735
   x 4

(example grid)
a. 7 2
  + 2 1
    9 3

Part 4 **Write the answers to the problems.**

a. $15 - 8 = \blacksquare$ b. $14 - 7 = \blacksquare$ c. $15 - 7 = \blacksquare$ d. $16 - 7 = \blacksquare$

e. $14 - 6 = \blacksquare$ f. $14 - 7 = \blacksquare$ g. $14 - 9 = \blacksquare$ h. $15 - 9 = \blacksquare$

i. $15 - 7 = \blacksquare$ j. $17 - 9 = \blacksquare$ k. $17 - 8 = \blacksquare$ l. $17 - 7 = \blacksquare$

Part 5

	Pond A	Pond B	Total for both ponds
Big fish			
Small fish	200		640
Total fish			780

Fact 1: There are 50 big fish in pond B.

Fact 2: In pond A, there are 110 more small fish than big fish.

a. How many big fish are in Pond B?

b. Are there more small fish in Pond A or in Pond B?

c. What's the total number of fish in Pond A?

d. Which Pond has more big fish?

e. If you wanted to go fishing for big fish, which Pond would you fish in?

3   6   9
→

12   15   18
→

21   24   27
→

30
→

Independent Work

Part 7 For some of these problems you have to carry. Copy each problem and figure out the answer.

a.
```
  7 2
+ 2 1
  9 3
```

a.   47
    x 5

b. 300
    x 2

c.   63
    x 2

d.   24
    x 9

Part 8 Copy the problem. Complete the sign.

a. $\dfrac{7}{4}$ ___ $\dfrac{6}{4}$

b. $\dfrac{3}{3}$ ___ 1

c. $\dfrac{2}{2}$ ___ $\dfrac{3}{2}$

d. $\dfrac{6}{6}$ ___ $\dfrac{4}{4}$

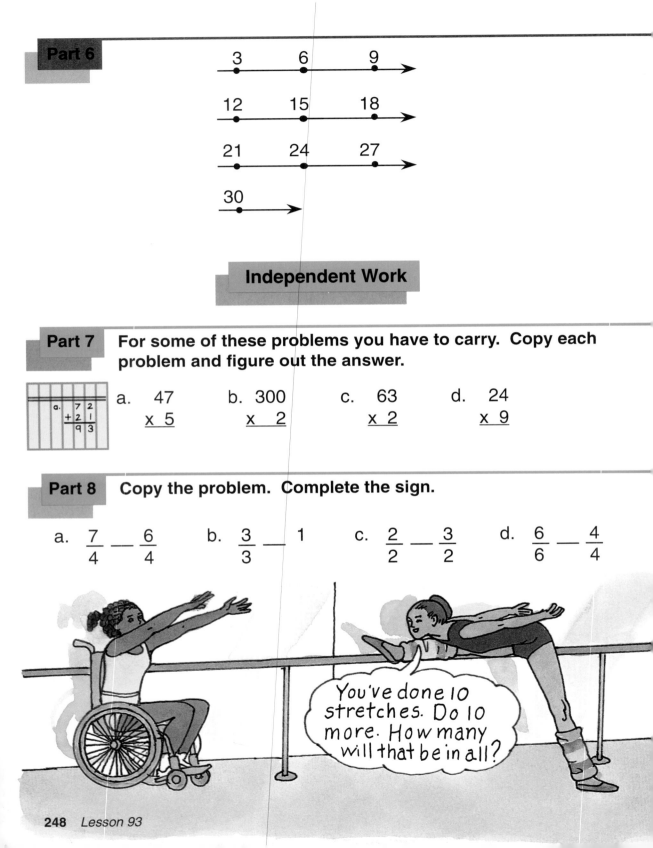

You've done 10 stretches. Do 10 more. How many will that be in all?

Write the number problems and the answers. Remember the unit names.

a. Seth was $\frac{7}{2}$ feet tall. Seth grew $\frac{3}{2}$ feet. How tall did Seth end up being?

b. A truck carried $\frac{4}{3}$ tons of stone. It delivered $\frac{1}{3}$ ton of stone. How much stone did the truck end up with?

c. A man had $\frac{5}{7}$ bag of concrete. He bought $\frac{14}{7}$ bags of concrete. How many bags of concrete did the man end up with?

d. A rope was $\frac{19}{2}$ yards long. $\frac{8}{2}$ yards were cut off the rope. How much rope was left?

Part 10 Make a number family for each problem. Then write the addition or subtraction problem for each family and figure out the answer. Remember the unit name.

a. A pig weighed 839 pounds. The pig lost some weight. The pig ended up weighing 803 pounds. How much weight did the pig lose?

b. A plane travelled 486 miles farther than a car. The car travelled 508 miles. How far did the plane travel?

c. A plane travelled 486 miles farther than a car. The plane travelled 808 miles. How far did the car travel?

d. A truck was filled with coal. The truck dumped 2018 pounds of coal. The truck ended up with 1000 pounds of coal. How many pounds of coal did the truck start out with?

Lesson 94

Part 1

- For each sentence you'll write two letters and a number in a multiplication number family.

- The word that follows the word **each** in these sentences tells about a small number.

- Here's a sentence: **Each <u>dime</u> is worth 10 cents.**

- Here's how to make the number family:

$$10 \overline{} \xrightarrow{\quad D \quad} C$$

- 10 and **dimes** are the small numbers. **Cents** is the big number.

Make a number family for each sentence.

a. Each <u>quarter</u> is worth 25 cents.

b. Each <u>dollar</u> is worth 100 cents.

c. Each <u>glerm</u> is worth 17 cents.

d. Each <u>ticket</u> is worth 9 dollars.

Part 2 Write the answers to the problems.

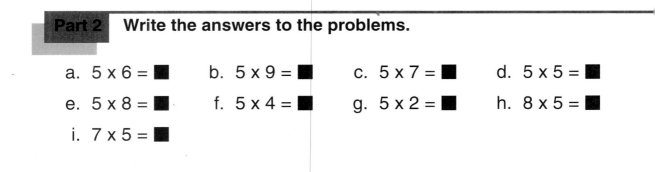

a. 5 x 6 = ■   b. 5 x 9 = ■   c. 5 x 7 = ■   d. 5 x 5 = ■

e. 5 x 8 = ■   f. 5 x 4 = ■   g. 5 x 2 = ■   h. 8 x 5 = ■

i. 7 x 5 = ■

Write the number families for the problems and figure out the answers.

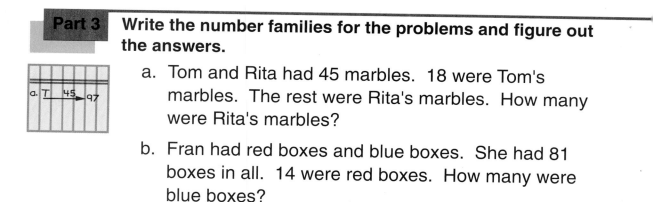

a. Tom and Rita had 45 marbles. 18 were Tom's marbles. The rest were Rita's marbles. How many were Rita's marbles?

b. Fran had red boxes and blue boxes. She had 81 boxes in all. 14 were red boxes. How many were blue boxes?

c. A bug had red spots and blue spots. The bug had 56 blue spots. The bug had 90 spots in all. How many red spots did the bug have?

Part 4

$$3 \qquad 6 \qquad 9 \longrightarrow$$

$$12 \qquad 15 \qquad 18 \longrightarrow$$

$$21 \qquad 24 \qquad 27 \longrightarrow$$

$$30 \longrightarrow$$

Independent Work

Part 5 **Copy the problems you know how to work. Then write the answers.**

a. $\dfrac{7}{4} + \dfrac{7}{3} =$ b. $\dfrac{8}{8} - \dfrac{2}{8} =$ c. $\dfrac{16}{4} + \dfrac{4}{4} =$ d. $\dfrac{4}{6} - \dfrac{4}{6} =$

Part 6   Write the answer to each problem.

a. 16 − 7 = ■   b. 15 − 7 = ■   c. 14 − 7 = ■   d. 15 − 8 = ■

e. 14 − 8 = ■   f. 15 − 9 = ■   g. 13 − 7 = ■   h. 13 − 8 = ■

i. 13 − 9 = ■   j. 15 − 7 = ■   k. 14 − 6 = ■   l. 13 − 5 = ■

Part 7   Write the addition problems and answers with a dollar sign.

	1	2	3	4	5
a. 7 2 / + 2 1 / 9 3	$6.13	$ .96	$3.08	$5.04	$5.86

a. A boy buys items 1 and 5. **Exactly** how much does he spend?

b. A girl buys items 1, 2 and 3. **About** how much does she spend?

Part 8   For each family write the multiplication fact that starts with the first small number.

a. 4 ⎯⎯▶ 24   b. 5 ⎯⎯▶ 20   c. 2 ⎯⎯▶ 16   d. 4 ⎯⎯▶ 36

e. 5 ⎯⎯▶ 15   f. 4 ⎯⎯▶ 28   g. 4 ⎯⎯▶ 32   h. 5 ⎯⎯▶ 10

i. 4 ⎯⎯▶ 12   j. 4 ⎯⎯▶ 36

Part 9   Copy each problem and work it.

a. 38
   × 2

b. 67
   + 5

c. 84
   − 8

d. 59
   × 6

e. 904
   − 852

f. 107
   + 93

g. 32
   × 4

Lesson 95

Part 1

- The problem is 36 plus 51.
- Below the problem is an arrow that shows the keys you press to get the answer.

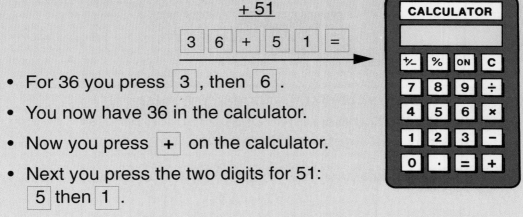

$$\begin{array}{r} 36 \\ + 51 \\ \hline \end{array}$$

- For 36 you press 3 , then 6 .
- You now have 36 in the calculator.
- Now you press + on the calculator.
- Next you press the two digits for 51: 5 then 1 .
- The last key is = . When you press that key, you get the answer.

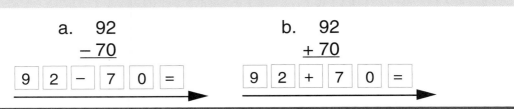

$$\begin{array}{r} a. \quad 92 \\ -70 \\ \hline \end{array} \qquad \begin{array}{r} b. \quad 92 \\ +70 \\ \hline \end{array}$$

Part 2 Make a number family for each sentence.

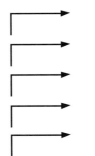

a. Each ticket is worth 7 dollars.

b. Each rop is worth 13 grapes.

c. Each glink is worth 71 bananas.

d. Each wup is worth 45 apples.

e. Each bug is worth 8 cents.

a. 5 x 6 = ■   b. 5 x 8 = ■   c. 5 x 10 = ■   d. 5 x 9 = ■

e. 5 x 5 = ■   f. 6 x 5 = ■   g. 7 x 5 = ■   h. 9 x 5 = ■

i. 8 x 5 = ■   j. 4 x 5 = ■   k. 2 x 5 = ■   l. 3 x 5 = ■

m. 5 x 7 = ■   n. 5 x 9 = ■   o. 5 x 8 = ■

Independent Work

Part 4 **Make a number family for each problem.
Then write the addition or subtraction problem for each
family and figure out the answer.**

a. A ranch had 451 horses. Some were wild and some
   were tame. There were 239 tame horses. How many
   were wild?

b. Ginger had old stamps and new stamps. If she had 33
   old stamps and 129 new stamps, how many stamps did
   she have in all?

c. There were 137 bugs in a barn. 78 were flies. The rest
   were not flies. How many of the bugs were not flies?

Part 5 **Write the answer to each problem.**

a. 17 – 8 = ■   b. 16 – 7 = ■   c. 15 – 8 = ■   d. 13 – 7 = ■

e. 15 – 7 = ■   f. 17 – 7 = ■   g. 11 – 7 = ■   h. 14 – 7 = ■

i. 11 – 6 = ■   j. 11 – 8 = ■   k. 11 – 10 = ■   l. 15 – 8 = ■

Part 6 Copy the table. Use the facts and figure out all the numbers for the table. Then answer the questions.

This table is supposed to show the bugs and frogs in Pond A and Pond B.

	Pond A	Pond B	Total for both ponds
Bugs			219
Frogs			
Total animals			

Fact 1: There are 150 bugs in Pond B.

Fact 2: There are 342 animals in Pond A.

Fact 3: The number of frogs in Pond A is 273.

Fact 4: There are 232 more frogs in Pond B than there are in Pond A.

a. Are there more animals in Pond A or in Pond B?

b. How many bugs are there?

c. How many animals are there?

d. In Pond A are there more bugs or frogs?

Part 7 Write the problems and the answers. Remember the unit name.

a. Joe had $\frac{10}{3}$ bags of chips. He ate $\frac{7}{3}$ bags of chips. How many bags did Joe end up with?

b. A cat weighed $\frac{19}{2}$ pounds. The cat gained $\frac{3}{2}$ pounds. How much did the cat end up weighing?

c. An elephant had $\frac{7}{4}$ gallons of water in his trunk. The elephant sucked $\frac{2}{4}$ more gallon into his trunk. How many gallons of water did the elephant have in his trunk?

Do the independent work for Lesson 95 in your workbook.

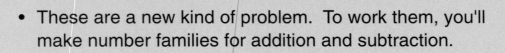

Part 1

- These are a new kind of problem. To work them, you'll make number families for addition and subtraction.

- Here's a problem: **Juan had quarters and dimes. He had 50 cents in quarters. He had 80 cents in all. How many cents did he have in dimes?**

- The word that tells about all the things in the problem is **cents.**

- The word **cents** goes under the arrow.

- The small numbers on the arrow are **Q** for quarters and **D** for dimes.

- Then you put in the numbers you know: The cents for quarters and the cents Juan had in all.

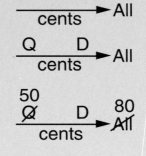

a. Tom had nickels and dimes. Tom had 40 cents in nickels. Tom had 70 cents in dimes. How many cents did Tom have in all?

b. Alice had pennies and quarters. She had 20 cents in pennies and 70 cents in all. How many cents did she have in quarters?

c. Rico had nickels and dimes. Rico had 20 cents in nickels. He had 50 cents in all. How many cents did he have in dimes?

d. Nita had pennies and dimes. Nita had 17 cents in pennies. She had 30 cents in dimes. How many cents did she have in all?

Part 2 — Work these problems on your calculator.

a.
$$57 + 63$$

| 5 | 7 | + | 6 | 3 | = | → |

b.
$$48 + 90$$

| 4 | 8 | + | 9 | 0 | = | → |

c.
$$70 - 44$$

| 7 | 0 | – | 4 | 4 | = | → |

Part 3 — Make a number family for each sentence.

a. Each hoop is worth 9 bananas.
b. Each room contains 9 people.
c. Each crate contained 34 oranges.
d. Each yard contained 4 gardens.
e. Each cat had 27 fleas.

Part 4 — Paired Practice

a. $16 - 7 = \blacksquare$ b. $17 - 10 = \blacksquare$ c. $15 - 8 = \blacksquare$ d. $14 - 6 = \blacksquare$

e. $14 - 7 = \blacksquare$ f. $15 - 7 = \blacksquare$ g. $16 - 7 = \blacksquare$ h. $13 - 6 = \blacksquare$

i. $13 - 8 = \blacksquare$ j. $15 - 6 = \blacksquare$ k. $17 - 8 = \blacksquare$ l. $15 - 8 = \blacksquare$

m. $13 - 6 = \blacksquare$ n. $11 - 8 = \blacksquare$ o. $11 - 5 = \blacksquare$

Part 5 Copy the problems you can work. Then write the answers.

a. $\dfrac{12}{5} - \dfrac{9}{5} =$

b. $\dfrac{12}{9} - \dfrac{12}{3} =$

c. $\dfrac{3}{4} + \dfrac{3}{3} =$

d. $\dfrac{8}{8} + \dfrac{5}{8} =$

e. $\dfrac{10}{3} - \dfrac{9}{9} =$

f. $\dfrac{8}{7} + \dfrac{10}{7} =$

Part 6 Make a number family for each problem. Then write the addition or subtraction problem for each family and figure out the answer. Remember the unit name.

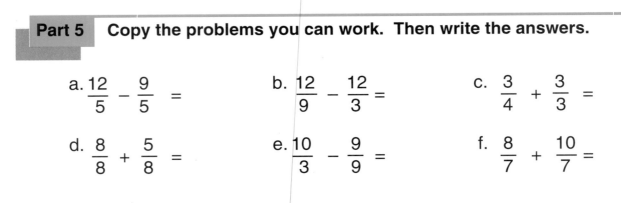

a. Joe worked 107 minutes longer than Debbie. Joe worked 691 minutes. How long did Debbie work?

b. A rabbit ran 186 yards from its den. Then the rabbit hopped farther away from its den. The rabbit ended up 254 yards away from its den. How many yards did the rabbit hop?

c. A girl had some cookies. She ate 97 cookies. She ended up with 97 cookies. How many cookies did the girl start out with?

d. A cow weighed 488 pounds less than a bull. The cow weighed 1207 pounds. How much did the bull weigh?

Lesson 97

Part 1

- In this multiplication problem you carry twice.
- First you work the ones, then the tens, then the hundreds.

$$\begin{array}{r} 1\ 3 \\ 5\ 3\ 7 \\ \times \quad 5 \\ \hline 2\ 6\ 8\ 5 \end{array}$$

a. $\begin{array}{r} 987 \\ \times \quad 5 \\ \hline \end{array}$
b. $\begin{array}{r} 738 \\ \times \quad 2 \\ \hline \end{array}$
c. $\begin{array}{r} 459 \\ \times \quad 4 \\ \hline \end{array}$
d. $\begin{array}{r} 594 \\ \times \quad 5 \\ \hline \end{array}$

Part 2

Find the sentence that tells about each and make the number family.
Figure out the missing number.
Then answer the question the problem asks.
Write the number and the name the problem asks about.

a. Each box holds 7 cans. You have 35 cans. How many boxes do you have?

b. Each room had 10 lights. There were 8 rooms. How many lights were there in all the rooms?

c. Each dog had 9 bugs. There were 7 dogs. How many bugs were there on all the dogs?

d. Each cat had 4 fleas. There were 36 fleas in all. How many cats were there?

e. Each boy ate 2 hot dogs. There were 5 boys. How many hot dogs were eaten by all the boys?

Work these problems on your calculator.

a.   75
   − 56

b.   38
   + 98

c.   54
   − 45

Make a number family for each problem.
Write the word below the arrow that tells about all the things.
Write the letters in the family and put in the numbers you know.
Then write the number problem and the answer. Remember the
unit name.

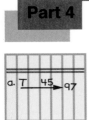

a. Jan had dimes and quarters. She had 60 cents in dimes and
75 cents in quarters. How many cents did she have in all?

b. Fran had dimes and nickels. She had 130 cents in all. She had
40 cents in nickels. How many cents did she have in dimes?

c. Frank had pennies and nickels. He had 98 cents in all. He had
33 cents in pennies. How many cents did he have in nickels?

Paired Practice

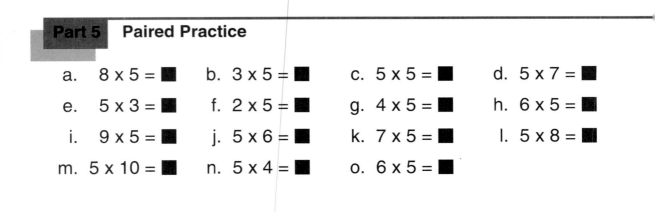

a. $8 \times 5 = \blacksquare$ b. $3 \times 5 = \blacksquare$ c. $5 \times 5 = \blacksquare$ d. $5 \times 7 = \blacksquare$

e. $5 \times 3 = \blacksquare$ f. $2 \times 5 = \blacksquare$ g. $4 \times 5 = \blacksquare$ h. $6 \times 5 = \blacksquare$

i. $9 \times 5 = \blacksquare$ j. $5 \times 6 = \blacksquare$ k. $7 \times 5 = \blacksquare$ l. $5 \times 8 = \blacksquare$

m. $5 \times 10 = \blacksquare$ n. $5 \times 4 = \blacksquare$ o. $6 \times 5 = \blacksquare$

Part 6 Write the problems and the answers. Remember the unit name.

a. A goat ate $\frac{7}{3}$ carrots. The goat ate $\frac{4}{3}$ more carrots. How many carrots did the goat eat in all?

b. A rope was $\frac{10}{2}$ yards long. If $\frac{3}{2}$ yards of the rope is cut off, how much of the rope is left?

c. There is $\frac{3}{4}$ gallon of milk in a refrigerator. A person puts $\frac{6}{4}$ more gallons of milk in the refrigerator. How much milk is in the the refrigerator?

Part 7 Write the addition problems and answers with a dollar sign.

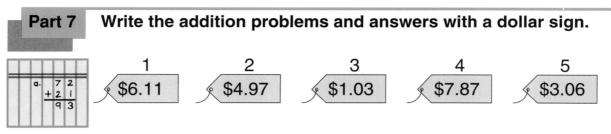

| 1 | 2 | 3 | 4 | 5 |
| $6.11 | $4.97 | $1.03 | $7.87 | $3.06 |

a. A girl buys items 2, 4 and 5. **About** how much does she spend?

b. A girl buys items 2, 4 and 5. **Exactly** how much does she spend?

c. A woman buys items 1, 2 and 3. **About** how much does she spend?

d. A boy buys items 1, 2 and 4. **Exactly** how much does he spend?

Part 8 Copy each problem. Complete the sign.

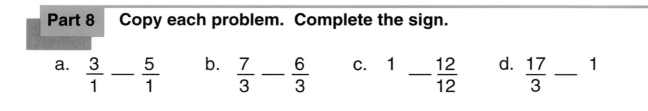

a. $\frac{3}{1}$ __ $\frac{5}{1}$ b. $\frac{7}{3}$ __ $\frac{6}{3}$ c. 1 __ $\frac{12}{12}$ d. $\frac{17}{3}$ __ 1

Write the answer to each problem.

a. 11
− 7

b. 13
− 6

c. 15
− 7

d. 9
− 5

e. 12
− 7

f. 16
− 6

g. 14
− 8

h. 12
− 5

i. 13
− 8

j. 15
− 8

k. 14
− 6

Part 10 Copy the table. Use the facts and figure out all the missing numbers for the table. Then answer the questions.

This table is supposed to show the dogs and cats that were in Pine Valley and Oak Park.

	Pine Valley	Oak Park	Total for both places
Dogs			184
Cats			
Pets			

Fact 1: There were 72 cats in Oak Park.

Fact 2: There were 126 cats in Pine Valley.

Fact 3: There were 382 pets in both places.

Fact 4: The number of cats in Oak Park was 22 less than the number of dogs in Oak Park.

a. Were there fewer cats or fewer dogs in both places?

b. How many dogs were in Pine Valley?

c. Were there more pets in Pine Valley or in Oak Park?

d. How many cats were there in both places?

Part 1

Work each problem with the calculator.
If the answer you get is the same as the answer shown,
write **ok.** If you get an answer that's different from the one
shown, write the correct answer.

a.
$$\begin{array}{r} 52 \\ + 68 \\ \hline 120 \end{array}$$

b.
$$\begin{array}{r} 703 \\ - 54 \\ \hline 549 \end{array}$$

c.
$$\begin{array}{r} 89 \\ - 41 \\ \hline 32 \end{array}$$

d.
$$\begin{array}{r} 230 \\ + 586 \\ \hline 826 \end{array}$$

e.
$$\begin{array}{r} 53 \\ - 39 \\ \hline 26 \end{array}$$

Part 2

Read the problem. Find the sentence that tells about
each. Make the number family for the problem. Figure out
the missing number. Then answer the question the
problem asks.

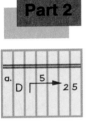

a. There were 9 rooms. Each room had 15 lights. How
many lights were there?

b. Each row had 4 squares. There were 24 rows. How
many squares were there in all?

c. There were 72 squares in all. Each row had 9 squares.
How many rows were there?

d. There were 5 rows. Each row had 20 squares. How
many squares were there in all?

- Some problems use a word that tells about all the things in the problem. For those problems you write that word under the number family.

- Some problems tell what happened first and what happens next. For those problems you put in the values forward or backward in the number family.

- Some problems tell about who had more or who had less. For those problems you put a letter in the number family for who has more and who has less.

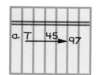

a. 36 bottles were empty. The rest of the bottles were full. There were 300 bottles in all. How many bottles were full?

b. First, Jane walked 13 miles. Then she rode a bike for some more miles. She travelled 41 miles on the whole trip. How many miles did she go on the bike?

c. The lion weighed 120 pounds less than the tiger. The tiger weighed 703 pounds. How many pounds did the lion weigh?

d. River City is 34 miles closer than Grove City. River City is 17 miles away. How many miles away is Grove City?

Paired Practice

a. 8 x 5 = ■   b. 9 x 5 = ■   c. 5 x 5 = ■   d. 6 x 5 = ■

e. 5 x 4 = ■   f. 5 x 6 = ■   g. 5 x 8 = ■   h. 5 x 3 = ■

i. 5 x 7 = ■   j. 5 x 9 = ■   k. 5 x 10 = ■   l. 5 x 3 = ■

m. 4 x 5 = ■   n. 8 x 5 = ■   o. 6 x 5 = ■   p. 7 x 5 = ■

Independent Work

Part 5   Write what X equals and what Y equals for each letter.

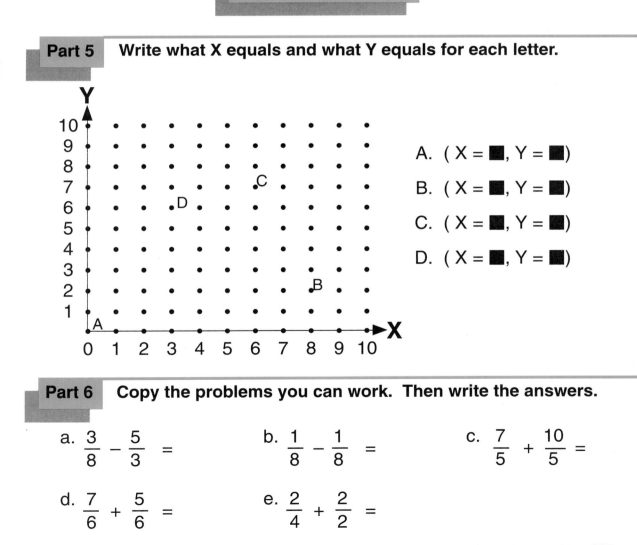

A. (X = ■, Y = ■)

B. (X = ■, Y = ■)

C. (X = ■, Y = ■)

D. (X = ■, Y = ■)

Part 6   Copy the problems you can work. Then write the answers.

a. $\dfrac{3}{8} - \dfrac{5}{3} =$

b. $\dfrac{1}{8} - \dfrac{1}{8} =$

c. $\dfrac{7}{5} + \dfrac{10}{5} =$

d. $\dfrac{7}{6} + \dfrac{5}{6} =$

e. $\dfrac{2}{4} + \dfrac{2}{2}$

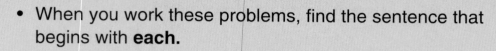

Part 1

- When you work these problems, find the sentence that begins with **each.**

- That sentence gives you information about making a multiplication number family.

- The name after **each** is a small number. The other name is the big number.

a. There were 8 rooms. Each room had 9 tables. How many tables were there in all?

b. There were 14 girls. Each girl had 9 cents. How many cents were there in all?

c. Each dog had 5 fleas. There were 45 fleas in all. How many dogs were there?

d. Each dog had 36 fleas. There were 4 dogs. How many fleas were there in all?

e. There were 28 plates in all. Each table had 4 plates on it. How many tables were there?

Part 2   **Work these problems on your calculator.**

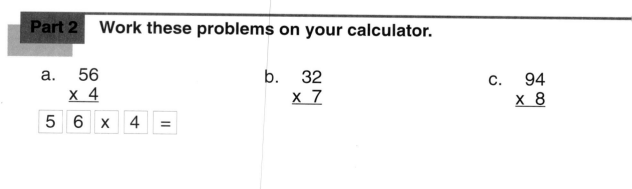

a. 56
   x 4

 5 6 x 4 =

b. 32
   x 7

c. 94
   x 8

a. The dog was 45 pounds heavier than the cat. The cat weighed 11 pounds. How many pounds did the dog weigh?

b. On Wednesday, Dan chopped down 16 trees. On Thursday, he chopped down some more trees. He chopped a total of 51 trees on both days. How many trees did he chop down on Thursday?

c. Dan had some seashells. Then he bought 123 seashells. He ended up with a total of 145 seashells. How many seashells did he start out with?

d. Park A had 34 more pine trees than Park B. Park A had 456 pine trees. How many pine trees did Park B have?

Independent Work

Part 4 **Copy each problem and work it.**

a. 785	b. 513	c. 985	d. 808	e. 386
− 697	− 96	+ 606	− 415	+ 308

Part 5 **Work each problem and write the answer. Remember the unit name.**

a. Joe had $\frac{11}{6}$ pies. He bought $\frac{9}{6}$ more pies. How many pies did Joe end up with?

b. A horse weighs $\frac{2}{5}$ tons. A rhino weighs $\frac{17}{5}$ tons. How much do the rhino and the horse weigh together?

c. A rat weighs $\frac{11}{3}$ pounds less than a cat. A cat weighs $\frac{15}{3}$ pounds. How much does the rat weigh?

Part 1

- For these problems, the question tells which name is the big number.

- Here's a problem: **A pine tree was 160 years old. A redwood tree was 450 years old. How much older was the redwood than the pine?**

- The question tells which tree was older.

- The redwood is the big number, and the pine is a small number.

- Here's the number family: $\xrightarrow{\quad P\quad} R$

- The question asks, "How much older?" So you write a box for the other small number. $\square \xrightarrow{\quad P\quad} R$

- Now you can put in the numbers you know and work the problem.

a. Rock Road is 89 miles long. Crow Road is 19 miles long. How many miles longer is Rock Road than Crow Road?

b. A train travelled 300 miles. A plane travelled 180 miles. How much farther did the train travel than the plane travelled?

Part 2   Make the number family for each problem. Figure out the answer to the question the problem asks.

a. Each dog had 3 fleas. There were 27 fleas in all. How many dogs were there?

b. There were 4 rooms. Each room had 12 windows. How many windows were there in all?

c. Each room had 5 tables. There were 20 tables in all. How many rooms were there?

d. There were 91 bugs. Each bug had 6 legs. How many legs were there?

e. Each bug had 6 legs. There were 12 legs in all. How many bugs were there?

Independent Work

Part 3 Write the addition problems and answers with a dollar sign.

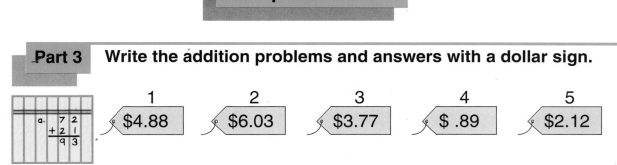

1	2	3	4	5
$4.88	$6.03	$3.77	$ .89	$2.12

a. A woman buys items 1, 2 and 4. **Exactly** how much does she spend?

b. A man buys items 2, 3 and 4. **About** how much does he spend?

c. A girl buys items 1, 3 and 5. **About** how much does she spend?

d. A boy buys items 2 and 5. **Exactly** how much does he spend?

Part 4 Copy each problem and work it.

a. 647   b.   42   c. 147   d.   59   e. 243
   x   5   x   3   x   9   x   6   x   2

Part 5 Copy each problem and work it.

a.   243   b.   642   c.   482   d.   505
   − 187   + 308   9   − 57
                          + 316

Part 6 — Write the time for each clock.

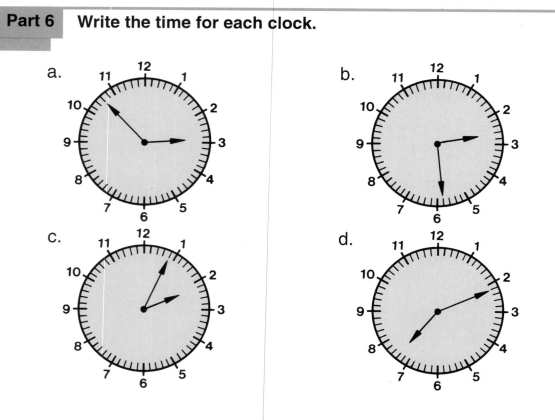

a.

b.

c.

d.

Part 7

Make the multiplication number family for each problem.
Put in the numbers that the problem gives you.
Find the missing number and answer the question.

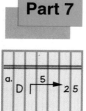

a. Sam has dimes. He has 90 cents in all. How many dimes does Sam have?

b. A girl has nickels. She has 8 nickels. How many cents does the girl have in all?

c. Debbie has nickels. She has 25 cents in nickels. How many nickels does Debbie have?

Do the independent work for lesson 100 in your workbook.

Test 10

Part 5 Work each problem and write the answer. Remember the unit name.

a. A bird weighed $\frac{6}{3}$ pounds. Then the bird lost $\frac{1}{3}$ pound. How many pounds did the bird end up weighing?

b. A bag of nails weighed $\frac{7}{4}$ pounds. Somebody put $\frac{5}{4}$ more pounds of nails in the bag. How many pounds were in the bag?

Part 6 For each item, make a number family, write the number problem and figure out the answer.

a. There were 8 rooms. Each room had 9 tables. How many tables were there in all?

b. There were 14 children. Each child had 9 pennies. How many pennies were there in all?

Part 7 Make a number family for each problem. Then write the number problems and figure out the answers. Remember the unit names.

a. There were 60 cars in all. 14 were red cars. The rest were blue cars. How many blue cars were there?

b. Tim had red marbles and green marbles. 16 were green marbles. 14 were red marbles. How many marbles did he have in all?

Lesson 101

Part 1   Use the numbers in the table to answer the questions.

- These questions ask you to compare two things.
- You can answer the questions by making number families with two letters.
- Then you put the numbers that are shown in the table into the number families.

This table shows the number of blue birds and red birds that were in Rock Park and Creek Park.

	Blue birds	Red birds	Total birds
Rock Park	153	8	161
Creek Park	137	206	343
Total for both parks	290	214	504

a. How many more blue birds were in Rock Park than in Creek Park?

b. In Rock Park, how many more blue birds were there than red birds?

c. How many more birds were in Creek Park than in Rock Park?

Part 2

a. $4 \overline{)36}$ b. $2 \overline{)14}$ c. $4 \overline{)28}$

d. $7 \overline{)63}$ e. $4 \overline{)24}$ f. $5 \overline{)45}$

Part 3   Copy each problem and work it. Then use your calculator to check your answers.

a.
$$\begin{array}{r} 97 \\ \times\ 4 \\ \hline \end{array}$$

b.
$$\begin{array}{r} 62 \\ \times\ 5 \\ \hline \end{array}$$

c.
$$\begin{array}{r} 74 \\ \times\ 9 \\ \hline \end{array}$$

d.
$$\begin{array}{r} 245 \\ \times\ 6 \\ \hline \end{array}$$

a.
$$\begin{array}{r} 7\ 2 \\ +\ 2\ 1 \\ \hline 9\ 3 \end{array}$$

Make a number family for each problem. Then write the addition problem or subtraction problem and the answer. Remember the unit name.

a. A redwood tree was 320 feet tall. A fir was 186 feet tall. How much taller was the redwood than the fir?

b. Ginger weighed 102 pounds. Frank weighed 148 pounds. How much heavier was Frank than Ginger?

c. Jill had 230 dollars. Fran had 89 dollars. How many more dollars did Jill have than Fran had?

d. A train travelled 394 miles. A car travelled 586 miles. How much farther did the car travel than the train?

Independent Work

Part 5 Write a number family for each problem and figure out the answer.

a. Each pack had 6 bottles. There were 30 bottles. How many packs were there?

b. Each dog has 4 paws. If there were 8 dogs, how many paws were there?

c. Joe had nickels. He had 45 cents in nickels. How many nickels did he have?

d. Tina had 6 dimes. How many cents in dimes did Tina have?

Part 6   Copy each problem and work it.

a.   24   b. 643   c. 514   d.   39
   x 9   x 2   x 6   x 4

Part 7   Copy each problem and work it.

a.   908   b.   389   c.   612   d.   808
   − 85   + 513   − 297   + 592

Part 8   Figure out the answer to each problem.
Write just the addition or subtraction problem for **a** and **b**.
Write the number family for problems **c** and **d.** Then write
the number problems.

a. A horse's tail was $\frac{5}{3}$ feet long. The tail grew $\frac{2}{3}$ of a foot.
How many feet long did the tail end up?

b. A pot had $\frac{9}{4}$ of a gallon in it. A boy took $\frac{3}{4}$ gallon out of the
pot. How many gallons were left in the pot?

c. Jan has 43 marbles. She has red marbles and green marbles.
How many red marbles does Jan have if 28 of the marbles are
green?

d. 474 people went to the fair. 238 of the people were females.
How many of the people were males?

Part 1

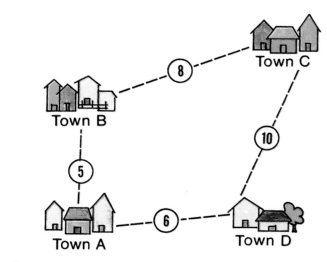

Use the map to answer the questions.

Problem 1: Carla and Jane went from town A to town C.
Carla said, "The trip was less than 14 miles."
Jane said, "The trip was more than 14 miles."

   a. Which person went through town B?

   b. How many miles did that person travel?

   c. Which person went through town D?

   d. How many miles did that person travel?

Problem 2: Don and Jackson went from town A to town D.
Don said, "The trip was less than 10 miles."
Jackson said, "The trip was more than 10 miles."

   a. Which person went through town B?

   b. How many miles did that person travel?

   c. Which person did not go through town B?

   d. How many miles did that person travel?

Write the whole number that equals each fraction.

a. $\blacksquare = \dfrac{30}{5}$ b. $\blacksquare = \dfrac{40}{5}$ c. $\blacksquare = \dfrac{54}{6}$

d. $\blacksquare = \dfrac{60}{6}$ e. $\blacksquare = \dfrac{27}{3}$ f. $\blacksquare = \dfrac{16}{2}$

Part 3   Use the numbers in the table to answer the questions.

- These questions ask you to compare two things.
- You can answer the questions by making number families with two letters.
- Then you put the numbers that are shown in the table into the number families.

This table shows the number of students in a school who play soccer or basketball.

	Boys	Girls	Total students
Soccer	180	220	400
Basketball	60	47	107
Total for both sports	240	267	507

a. How many more boys than girls play basketball?

b. How many more students play soccer than play basketball?

c. How many more girls play soccer than play basketball?

Part 4   Write the answer to each problem.

a. $5\overline{\smash{)}50}$ b. $5\overline{\smash{)}30}$ c. $5\overline{\smash{)}35}$ d. $5\overline{\smash{)}25}$ e. $5\overline{\smash{)}45}$

Part 5   Write the answer for each problem.

a. $2\overline{)16}$ b. $4\overline{)24}$ c. $3\overline{)15}$ d. $5\overline{)25}$

e. $2\overline{)18}$ f. $3\overline{)12}$ g. $8\overline{)72}$ h. $2\overline{)12}$

Part 6   Write a number family for each problem and figure out the answer. Remember the unit name.

a. Sue has 9 dimes. How many cents in all does Sue have?

b. Each can holds 3 ounces. There are 9 cans. How many ounces are there?

Part 7   Write the addition problems and answers with a dollar sign.

1	2	3	4	5
$6.87	$1.05	$4.10	$4.94	$5.92

a. The boy buys items 2 and 3. **About** how much does he spend?

b. A man buys items 1, 2 and 3. **Exactly** how much does he spend?

Part 8   Copy the problems you can work and work them.

a. $\dfrac{12}{5} - \dfrac{7}{5} =$ b. $\dfrac{8}{7} + \dfrac{2}{7} =$ c. $\dfrac{14}{8} + \dfrac{4}{8} =$

Part 9   Copy each problem and work it.

a. $387 + 5$ b. $463 - 378$ c. 317×5 d. $809 - 57$

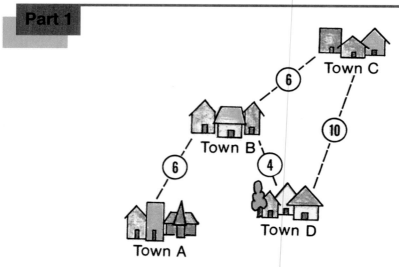

Use the map to answer the questions.

Problem 1: Fran and Frank went from town A to town C.
Fran said, "The trip was less than 15 miles."
Frank said, "The trip was more than 15 miles."

   a. Which person went through town D?

   b. How many miles did that person travel?

   c. Which person did not go through town D?

   d. How many miles did that person travel?

Problem 2: Bob and Barb went from town A to town D.
Bob said, "The trip was less than 15 miles."
Barb said, "The trip was more than 15 miles."

   a. Which person went through fewer towns?

   b. How many miles did that person travel?

   c. Which person went through more towns?

   d. How many miles did that person travel?

This table is supposed to show the number of men and women that ate at Kate's Cafe and Joe's Grill.

	Men	Women	Adults
Kate's Cafe		81	
Joe's Grill			
Total for both places			

Fact 1: 209 adults ate at Kate's Cafe.

Fact 2: The total men for both places was 213.

Fact 3: 94 women ate at Joe's Grill.

Fact 4: The number of adults who ate at Joe's Grill is 30 less than the number of adults who ate at Kate's Cafe.

a. How many more men ate at Kate's Cafe than at Joe's Grill?

b. How many more women ate at Joe's Grill than at Kate's Cafe?

c. How many more men than women ate at Kate's Cafe?

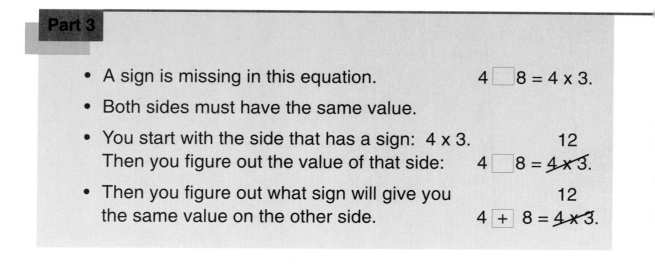

- A sign is missing in this equation. $4 \boxed{} 8 = 4 \times 3.$

- Both sides must have the same value.

- You start with the side that has a sign: 4×3.
  Then you figure out the value of that side: $4 \boxed{} 8 = \overset{12}{\cancel{4 \times 3}}.$

- Then you figure out what sign will give you the same value on the other side. $4 \boxed{+} 8 = \overset{12}{\cancel{4 \times 3}}.$

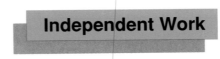

Part 4 Write the addition or subtraction problem and figure out the answer for each problem. Remember the unit name.

a. A tree was $\frac{5}{3}$ yards tall. A deer ate $\frac{3}{3}$ yard off the top of the tree. How tall did the tree end up?

b. $\frac{9}{4}$ pints was in a container. A woman drank $\frac{3}{4}$ of a pint. How many pints were left in the container?

c. A tree was $\frac{15}{2}$ feet tall. The tree grew $\frac{9}{2}$ feet more. How many feet tall did the tree end up?

Part 5 Copy each problem. Complete the sign.

a. $\frac{5}{4}$ ___ 1   b. $\frac{7}{7}$ ___ $\frac{9}{10}$ c. $\frac{8}{3}$ ___ $\frac{6}{3}$ d. 1 ___ $\frac{5}{7}$

Part 6 For some problems you have to carry and for some problems you don't have to carry. Copy each problem and figure out the answer.

a. 247
  x   5

b. 316
  x   2

c.   63
  x   2

d. 518
  x   5

Part 7 Use numbers close to the numbers in each problem to figure out **about** what the answer is.

a.   386
     98
  + 509

b.   878
   − 214

c.   506
   + 89

d. 1018
   − 319

Write a number family for each problem and figure out the answer.

a. Jim has 7 nickels. How many cents does Jim have?

b. Each dog has 4 paws. If there are 9 dogs, how many paws are there?

c. Each pack holds 6 cans. A store has 30 cans. How many packs does the store have?

d. Sue has 80 cents in dimes. How many dimes does Sue have?

Part 1

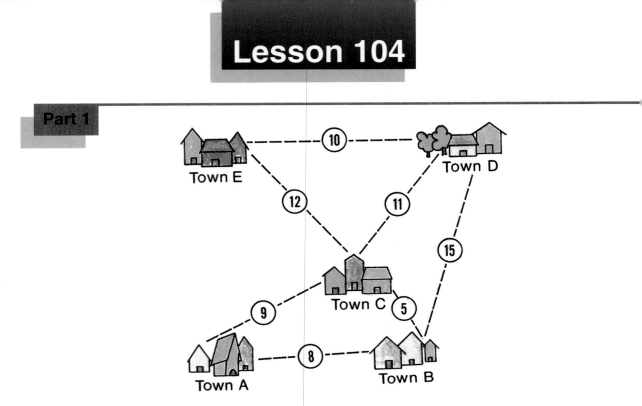

Problem 1: There are many ways to go from town A to town E.

    a. Figure out the route that is the longest and write the number of miles for that route.

    b. Figure out the route that is the shortest and write the number of miles for that route.

Problem 2: There are many ways to go from town A to town B.

    a. Figure out the longest route from town A to town B and write the number of miles for that route.

    b. Figure out the shortest route from town A to town B and write the number of miles for that route.

- These are just like problems you've worked, but for some of these problems, you'll have to write a column multiplication problem to figure out the answer.

a. Each bug laid 27 eggs. There were 5 bugs. How many eggs were laid in all?

b. Each row had 14 squares. There were 5 rows. How many squares were there in all?

c. Each dog had 3 fleas. There were 27 fleas in all. How many dogs were there?

d. Each room had 26 children. There were 3 rooms. How many children were there?

e. Each house had 8 doors. There were 72 doors in all. How many houses were there?

Independent Work

Part 3 Make a number family for each problem.
Then write the addition problem or subtraction problem and the answer. Remember the unit name.

a. Susan weighs 128 pounds less than Dan. Dan weighs 212 pounds. How much does Susan weigh?

b. A goat ran 390 yards farther than a sheep. The sheep ran 724 yards. How far did the goat run?

c. Joan is 26 inches shorter than Debbie. Debbie is 72 inches tall. How tall is Joan?

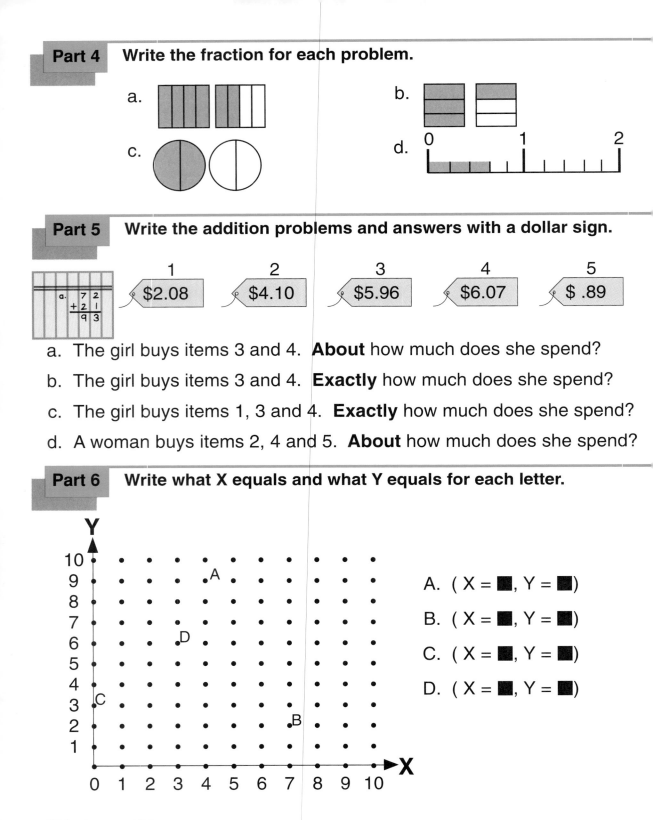

Part 4 Write the fraction for each problem.

a.

b.

c.

d. 0   1   2

Part 5 Write the addition problems and answers with a dollar sign.

a. 7 2
   + 2 1
   ——
   9 3

1 $2.08   2 $4.10   3 $5.96   4 $6.07   5 $ .89

a. The girl buys items 3 and 4. **About** how much does she spend?

b. The girl buys items 3 and 4. **Exactly** how much does she spend?

c. The girl buys items 1, 3 and 4. **Exactly** how much does she spend?

d. A woman buys items 2, 4 and 5. **About** how much does she spend?

Part 6 Write what X equals and what Y equals for each letter.

A. (X = ■, Y = ■)

B. (X = ■, Y = ■)

C. (X = ■, Y = ■)

D. (X = ■, Y = ■)

Part 1

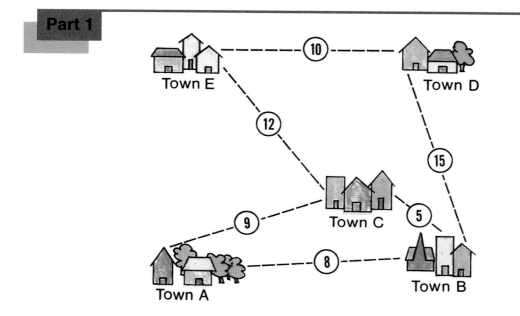

Problem 1: There are many ways to go from town A to town D.

    a. Figure out the route that is the longest and write the number of miles for that route.

    b. Figure out the route that is the shortest and write the number of miles for that route.

Problem 2: There are many ways to go from town A to town C.

    a. Figure out the longest route from town A to town C and write the number of miles for that route.

    b. Figure out the shortest route from town A to town C and write the number of miles for that route.

a. Each cow gave 5 gallons of milk. There were 30 cows. How many gallons of milk were there?

b. Each boy had 2 shoes. There were 20 shoes. How many boys were there?

c. Each barn held 15 cows. There were 9 barns. How many cows were there?

d. Each dog had 35 fleas. There were 3 dogs. How many fleas were there?

e. Each girl had 16 coins. There were 9 girls. How many coins were there?

Part 3

- Here's 4 times zero: $4 \times 0 = \blacksquare$.
- 4×0 tells that you have 4 zeros: $0 + 0 + 0 + 0 = 0$.
- Here's zero times 4: $0 \times 4 = \blacksquare$.
- Zero times 4 tells you how many fours you have.
- You have no fours: $4 \times 0 = 0$
  $$0 \times 4 = 0$$
- If you think of what a problem that multiplies by zero is telling you, you'll never get confused. If you multiply anything by zero, you end up with zero.
- But, remember, if you **add** or **subtract** zero, that's different. For those problems, you end up with the number you start out with.

$$5 - 0 = 5$$
$$7 + 0 = 7$$

Part 4   Copy the problems you can work. Then write the answers.

a. $\dfrac{15}{7} - \dfrac{15}{15} =$ b. $\dfrac{37}{2} + \dfrac{5}{2} =$ c. $\dfrac{3}{6} + \dfrac{42}{6} =$

d. $\dfrac{3}{4} + \dfrac{10}{4} =$ e. $\dfrac{5}{5} - \dfrac{5}{7} =$ f. $\dfrac{4}{1} + \dfrac{4}{3} =$

Part 5   Make a number family for each problem. Then write the addition problem or subtraction problem and the answer. Remember the unit name.

a. A girl has 95 cents in nickels and quarters. The girl has 20 cents in nickels. How many cents in quarters does the girl have?

b. A bowl contains red marbles and blue marbles. There are 37 blue marbles. If there are 124 marbles altogether, how many red marbles are in the bowl?

c. All the fish that live in a pond are black or gold. There are 65 black fish and 72 gold fish in the pond. How many fish live in the pond?

Part 6   Copy each problem and work it.

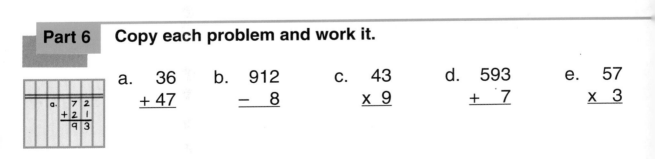

a.   36
   + 47

b.   912
   −   8

c.   43
   × 9

d.   593
   +   7

e.   57
   × 3

Part 7 — Make a number family for each problem and figure out the answer. Remember the unit name.

a. A man has dimes. A man has 90 cents in dimes. How many dimes does the man have?

b. Each bush had 25 flowers. There were 3 bushes. How many flowers were there?

c. Jill has 28 nickels. How many cents in nickels does Jill have?

d. There were 124 more green frogs than brown frogs in a pond. There were 779 green frogs. How many brown frogs were in the pond?

e. Donna had $4.60 less than Zetta. Zetta had $12.80. How much did Donna have?

I need 6 more paint brushes.

Lesson 106

Part 1

This table is supposed to show the daisies and roses in Bright Valley and Sunrise Estates.

	Bright Valley	Sunrise Estates	Total for both places
Daisies			
Roses			
Total flowers			749

Questions

a. How many more flowers are in Sunrise Estates than in Bright Valley?

b. How many more rose than daisies are in Sunrise Estates?

c. How many more daisies are in Sunrise Estates than in Bright Valley?

Fact 1: There are 245 roses in Sunrise estates.

Fact 2: There are 302 flowers in Bright Valley.

Fact 3: The number of roses in Sunrise Estates is 127 more than the number of roses in Bright Valley.

Fact 4: The number of daisies in Sunrise Estates is 43 less than the number of roses in Sunrise Estates.

Part 2

- You can read this fraction as a division problem: $\dfrac{27}{9}$

- 27 is the big number. 9 is a small number. The answer is the missing small number. $\dfrac{27}{9} = 3$

- Here's how you read it as a division fact: 27 divided by 9 equals 3.

Part 3 Each of these is missing an operation sign.
Copy each problem and complete the equation.

a. $17 - 5 = 19 \quad\blacksquare\quad 7$ b. $8 \quad\blacksquare\quad 3 + 1 = 12 - 0$ c. $20 + 8 = 32 \quad\blacksquare\quad 4$

Part 4 Each of these equations is missing a number.
Copy each problem and complete the equation.

a. $17 - 5 = 8 + \blacksquare$ b. $8 + 3 + 4 = 15 - \blacksquare$ c. $32 + 4 = 38 - \blacksquare$

Part 5 Write the addition or subtraction problem and figure out
the answer. Remember the unit name.

a. A plant was $\frac{10}{4}$ feet tall. The plant grew $\frac{3}{4}$ foot. How tall did
the plant end up?

b. A farmer had $\frac{14}{4}$ bags of food. The farmer fed his animals $\frac{4}{4}$
bag of food. How many bags of food did the farmer have left?

c. A carpenter had $\frac{11}{5}$ bags of nails. The carpenter bought $\frac{3}{5}$
bag of nails. How many bags did the carpenter end up with?

d. A man had $\frac{7}{5}$ gallons of paint. The man used up $\frac{7}{5}$ gallons of
paint. How much paint did the man end up with?

Part 6 Copy each problem. Complete the sign.

a. $1 \quad\underline{\quad}\quad \dfrac{5}{4}$ b. $\dfrac{9}{10} \quad\underline{\quad}\quad 1$ c. $\dfrac{8}{3} \quad\underline{\quad}\quad \dfrac{9}{3}$ d. $\dfrac{4}{4} \quad\underline{\quad}\quad 1$

Part 7 Copy each problem and work it.

a. $\begin{array}{r} 17 \\ \times\ 5 \\ \hline \end{array}$ b. $\begin{array}{r} 273 \\ \times\ \ 3 \\ \hline \end{array}$ c. $\begin{array}{r} 58 \\ \times\ 9 \\ \hline \end{array}$

Write the time for each clock.

a.

b.

c.

d.

Lesson 107

Part 1

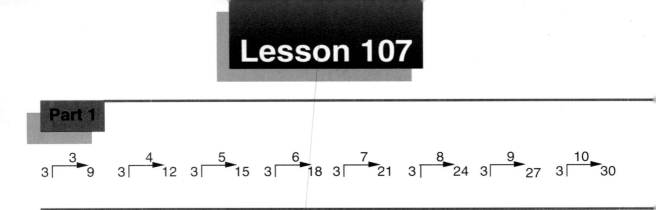

$$3\overline{\smash{\big)}\,9}^{\,3} \quad 3\overline{\smash{\big)}\,12}^{\,4} \quad 3\overline{\smash{\big)}\,15}^{\,5} \quad 3\overline{\smash{\big)}\,18}^{\,6} \quad 3\overline{\smash{\big)}\,21}^{\,7} \quad 3\overline{\smash{\big)}\,24}^{\,8} \quad 3\overline{\smash{\big)}\,27}^{\,9} \quad 3\overline{\smash{\big)}\,30}^{\,10}$$

Part 2

- To count out amounts like $2.35, always start with the **largest value** and get as close to the number as you can with that value.

- We want to get as close as we can to $2.35. So we start with the largest value–dollars.

- We circle 2 dollars.

- We circle 1 quarter.

- We circle 1 dime.

- We circled dollars and coins to show how we counted out $2.35.

$2.35

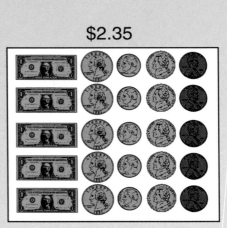

$2.35

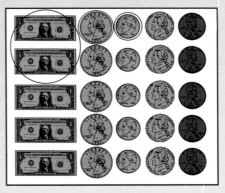

Part 3 Find the sentence that tells about **each** and make the number family. To work the problem write a column multiplication problem and figure out the answer.

a. A farmer has 42 goats. Each goat produces 3 pints of milk. How many pints of milk does the farmer get?

b. Each cake has 5 raisins. A baker makes 9 cakes. How many raisins does the baker use?

c. Each ride at the fair costs 70 cents. Jim has enough money to go on 4 rides. How many cents does he have?

Part 4 Copy the problems you can work. Then write the answers.

a. $\dfrac{20}{4} - \dfrac{20}{20} =$ c. $\dfrac{3}{2} - \dfrac{3}{2} =$ e. $\dfrac{9}{11} + \dfrac{2}{9} =$

b. $\dfrac{5}{7} + \dfrac{11}{7} =$ d. $\dfrac{6}{2} + \dfrac{6}{3} =$ f. $\dfrac{15}{10} - \dfrac{3}{10} =$

Part 5 Write the addition problems and answers with a dollar sign.

1	2	3	4	5
$2.08	$4.10	$5.96	$6.07	$ .89

a. The girl bought items 2 and 4. **About** how much did she spend?

b. The girl bought items 1 and 5. **About** how much did she spend?

c. A man bought items 1, 3 and 5. **About** how much did he spend?

d. A woman bought items 1, 2 and 4. **Exactly** how much did she spend?

Part 6

Write a number family for each problem.
Write the column problem and figure out the answer.
Remember the unit name.

a. Dan had many letters. He mailed 56 of the letters. He had 94 letters left. How many letters did Dan start out with?

b. A store had 378 more apples than pears. If the store had 543 apples, how many pears did the store have?

c. Mr. Brown had $295. Mrs. Smith had $532. How much less did Mr. Brown have than Mrs. Smith had?

d. A dog ran 624 yards. Then the dog walked some distance. The dog went 956 yards in all. How many yards did the dog walk?

Lesson 108

Part 1

a. $\dfrac{15}{3}$ b. $\dfrac{10}{2}$ c. $\dfrac{12}{3}$ d. $\dfrac{20}{4}$ e. $\dfrac{9}{3}$ f. $\dfrac{12}{2}$ g. $\dfrac{45}{5}$

Part 2

a. $2\overline{)18}$ b. $4\overline{)16}$ c. $6\overline{)18}$ d. $2\overline{)6}$ e. $5\overline{)45}$

f. $9\overline{)45}$ g. $3\overline{)9}$ h. $4\overline{)24}$ i. $4\overline{)32}$ j. $3\overline{)12}$

Part 3

- You've learned how to make multiplication number families for problems that tell about **each.**

- Here's a sentence: **Each room had 12 lights.**   12⌐——→ R L

- There are more lights than rooms. So **lights** is the big number.

- Here's the rule: The name that comes after the word **each** is the name of a small number.

- Here's the same sentence about the rooms and lights: **Each room** has 12 lights.

- The word after **each** is a small number. So **room** is a small number.

- Here's the information about the rooms and the lights written a different way: There were 12 lights in **each room.**

- The word after **each** is **room.** So **room** is still the small number. There are still more lights than rooms.   12⌐——→ R L

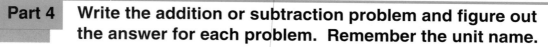

Part 4 Write the addition or subtraction problem and figure out the answer for each problem. Remember the unit name.

a. A kitten weighed $\frac{5}{3}$ pounds. The kitten gained $\frac{2}{3}$ pounds. How much did the kitten end up weighing?

b. A piece of string was $\frac{11}{3}$ feet long. $\frac{5}{3}$ feet is cut off of the string. How long does the string end up?

c. Donna has $\frac{3}{4}$ of a pie. She buys $\frac{6}{4}$ pies. How much pie does she end up with?

Part 5 Write the fraction for each item.

a.

b. 0   1   2

c. 0   1   2

d.

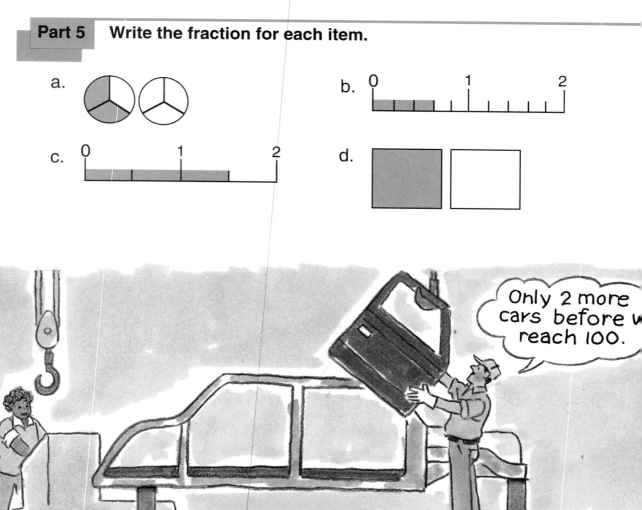

Only 2 more cars before w reach 100.

Part 6 Copy the table. Fill in the numbers. Write the answers to the questions.

This table is supposed to show the houses and sheds that have roofs made of metal and wood.

	Wood roofs	Metal roofs	Total for both roofs
Houses			
Sheds			
Total buildings			

Questions

a. How many more sheds than houses have metal roofs?

b. How many more buildings have metal roofs than have wood roofs?

c. How many more houses than sheds are there?

Fact 1: There are 97 houses with wood roofs.

Fact 2: The number of houses with metal roofs is 217 more than number of houses with wood roofs.

Fact 3: The number of houses with wood roofs is 89 more than number of sheds with wood roofs.

Fact 4: There are 324 sheds.

The body is 5 feet high. Bring it down 3 feet.

Lesson 109

Part 1

a. $17 - 9 = \blacksquare$ b. $17 - 0 = \blacksquare$ c. $17 \times 0 = \blacksquare$ d. $4 \times 6 = \blacksquare$

e. $4 \times 1 = \blacksquare$ f. $5 \times 1 = \blacksquare$ g. $5 \times 0 = \blacksquare$ h. $5 + 0 = \blacksquare$

i. $5 + 20 = \blacksquare$ j. $20 + 1 = \blacksquare$ k. $20 \times 1 = \blacksquare$ l. $20 \times 0 = \blacksquare$

Part 2

a. $6 \overline{) 18}$ b. $3 \overline{) 15}$ c. $3 \overline{) 12}$ d. $3 \overline{) 9}$ e. $3 \overline{) 6}$

f. $4 \overline{) 12}$ g. $3 \overline{) 18}$ h. $5 \overline{) 15}$ i. $4 \overline{) 24}$ j. $4 \overline{) 32}$

Part 3

- You're going to work division problems that have leftovers. But they don't divide by 10 or 5. They divide by 4.

- They work the same way the other problems work except that there is no **P** for the pennies. There's an **R.** That's the remainder. You just write the ones that are left over after the **R.**

- Here's a problem: $4 \overline{) 23}$

- 23 is not a big number for fours, so there are leftovers.

- You think of how many times you'd multiply by 4 to get as close as possible to 23.

- You write **5** above the second digit of 23: $4 \overline{) 2\,3}^{\,5}$

- Then you write what 4 times 5 equals just under 23, and you write the ones that are left over after the **R.**

$$\begin{array}{r} 5 \\ 4 \overline{) 2\,3} \\ \underline{2\,0} \ R3 \end{array}$$

Part 4   Copy each problem and complete the equation.

a. $18 - 3 = \boxed{} + 15$

b. $10 - 4 = 2 \boxed{} 3$

c. $30 \boxed{} 6 = 20 + 4$

d. $5 \times 4 = 22 - \boxed{}$

Part 5   Write a number family for each problem. Write a column problem and figure out the answer. Remember the unit name.

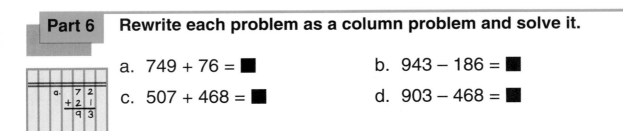

a. A store started with some cartons. The store sold 358 cartons. The store ended up with 803 cartons. How many cartons did the store start with?

b. Jim is 76 millimeters taller than Sue. Sue is 828 millimeters tall. How tall is Jim?

c. A tractor had 176 quarts of gasoline. Then some more gasoline was put in the tractor. The tractor ended up with 330 quarts. How many quarts were added?

Part 6   Rewrite each problem as a column problem and solve it.

a. $749 + 76 = \blacksquare$

b. $943 - 186 = \blacksquare$

c. $507 + 468 = \blacksquare$

d. $903 - 468 = \blacksquare$

Lesson 110

Part 1

- Some multiplication problems have a zero, and some problems add or subtract zero.

- Remember, when you multiply with zero the answer is always zero.

- When you add or subtract zero, you're not adding or subtracting anything, so the number you end up with is the number you start out with.

a. $30 + 50 =$ ■   f. $15 \times 1 =$ ■

b. $30 + 1 =$ ■   g. $15 - 1 =$ ■

c. $30 \times 1 =$ ■   h. $15 - 0 =$ ■

d. $30 \times 0 =$ ■   i. $15 + 1 =$ ■

e. $30 + 0 =$ ■   j. $15 \times 0 =$ ■

Write the answer for each problem.

a. 6 x 3 = ■   b. 4 x 3 = ■   c. 9 x 3 = ■   d. 3 x 8 = ■

e. 3 x 7 = ■   f. 3 x 6 = ■   g. 8 x 3 = ■   h. 3 x 9 = ■

i. 8 x 3 = ■   j. 3 x 10 = ■   k. 7 x 3 = ■   l. 6 x 3 = ■

Independent Work

Part 3 **Copy each problem and complete the equation.**

a. 15 − 6 = 3 ▨ 3   b. 17 ▨ 2 = 3 x 5

c. 25 ▨ 3 = 30 − 2   d. 38 + 1 = 40 − ▨

Part 4 **Copy each problem. Complete the sign.**

a. 1 __ $\frac{7}{8}$ b. $\frac{3}{7}$ __ $\frac{4}{7}$ c. $\frac{15}{14}$ __ 1   d. $\frac{25}{2}$ __ $\frac{3}{2}$

Part 5 **Write the addition or subtraction problem and figure out the answer. Remember the unit name.**

a. An elephant started out with $\frac{7}{3}$ buckets of food. The elephant ate $\frac{2}{3}$ bucket of the food. How many buckets of food did the elephant end up with?

b. A cat weighed $\frac{9}{5}$ pounds. The cat lost $\frac{3}{5}$ pounds. How much did the cat end up weighing?

c. A man talked on the phone for $\frac{7}{4}$ hours. He talked on the phone another $\frac{3}{4}$ hour. How many hours did he talk on the phone altogether?

Part 6 — Rewrite each problem as a column problem and solve it.

a. 74 x 2 = ■

b. 432 x 3 = ■

c. 316 x 4 = ■

d. 58 x 9 = ■

Part 7 — Make a number family for each problem. Then figure out the answer. Remember the unit name.

a. A deer weighed 457 pounds. An elk weighed 781 pounds. How much heavier was the elk than the deer?

b. A sports car could go 145 miles per hour. A race car could go 189 miles an hour. How many miles an hour faster was the race car then the sports car?

c. On Saturday, Greg spent $11.00. He spent some more money on Sunday. The total that he spent on both days was $24.02. How much money did he spend on Sunday?

d. The pole is 134 feet taller than the building. The building is 21 feet tall. How tall is the pole?

A person has a big bag of coins. The only coins in the bag are pennies, nickels and dimes. But there's lots of pennies, lots of nickels, and lots of dimes.

- Here's a problem: **A person reaches into the bag and pulls out 3 coins.** You have to figure out the **smallest** number of cents the person could have.

- Here's how you do that: You figure out which of the 3 coins is worth the least number of cents.

    If the person pulled out 3 pennies, the 3 coins would be worth only 3 cents. That's the smallest number of cents the person could have with 3 coins.

- You can also figure out the **largest** number of cents the person could have with 3 coins.

    Dimes are the coins that are worth the most. Each dime is worth 10 cents. If the person pulled out 3 dimes, the 3 coins would be worth 30 cents.

Write the answer to each question.

a. What's the smallest number of cents the person could have with 3 coins?

b. What's the largest number of cents the person could have with 3 coins?

c. If the person has 3 coins that are worth 7 cents, what are the 3 coins?

d. If the person has 3 coins that are worth 21 cents, what are the 3 coins?

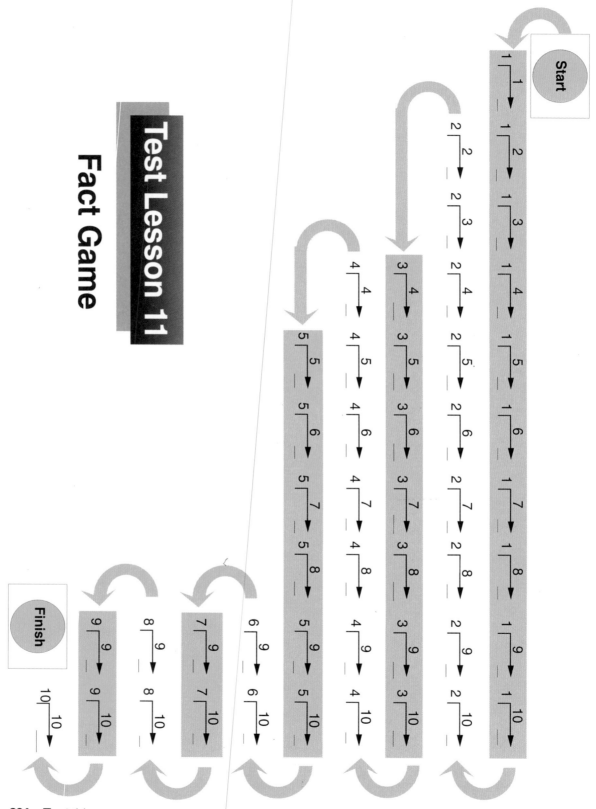

Test 11

Part 6

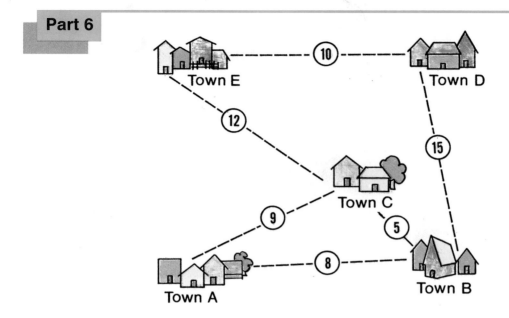

Use the map to figure the routes.

There are many ways to go from town C to town D.

a. Figure out the route that is the longest and write the number of miles for that route.
b. Figure out the route that is the shortest and write the number of miles for that route.

Part 7   **Write the answer.**

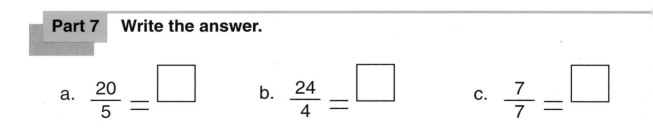

a. $\dfrac{20}{5} = \boxed{}$ b. $\dfrac{24}{4} = \boxed{}$ c. $\dfrac{7}{7} = \boxed{}$

Part 8 Copy and complete each equation.

a. $5 \times 8 = 50 \ \blacksquare \ 10$

b. $10 - 4 = 2 \ \blacksquare \ 3$

c. $30 \ \blacksquare \ 6 = 20 + 4$

d. $5 \times 4 = 22 - \blacksquare$

Part 9 Write the answers.

a. $45 \times 1 = \blacksquare$

b. $45 - 1 = \blacksquare$

c. $45 - 0 = \blacksquare$

d. $45 + 1 = \blacksquare$

e. $45 \times 0 = \blacksquare$

Part 10 Write the division problem and the answer.

a. $\dfrac{18}{2}$

b. $\dfrac{24}{4}$

Part 11 For each problem, find the sentence that tells about **each.** Make a number family for the problem and figure out the missing number. Then answer the question the problem asks.

a. Each dog had 3 fleas. There were 27 fleas in all. How many dogs were there?

b. There were 4 rooms. Each room had 12 fleas. How many fleas were there in all?

c. Each room had 5 tables. There were 50 tables in all. How many rooms were there?

d. There were 91 bugs. Each bug had 6 legs. How many legs were there?

Lesson 111

- Here's a sentence: **There were 9 bugs in each room.**

- The word after **each** is a small number.

- Here's the number family: $9 \overset{R}{\longrightarrow} B$

- When you think about the room and the bugs, you have to think about numbers, not size.

- The room is much bigger than a bug. But the **room** is a small number, not the big number. That's because the number of rooms is smaller than the number of bugs.

- When there's only 1 room, there's more than 1 bug. There are 9 bugs.

- When there are only 2 rooms, there are 18 bugs.

a. 5 bugs were in each can. There were 17 cans. How many bugs were there?

$D \overset{5}{\longrightarrow} 25$

b. They had 4 bugs in each net. There were 24 bugs. How many nets were there?

c. There were 13 bugs in each tree. There were 9 trees. How many bugs were there?

d. Each carton held 19 bugs. There were 2 cartons. How many bugs were there?

Part 2 Write each fraction as a division problem.
Then write the answers.

a. $\dfrac{20}{5}$ b. $\dfrac{16}{4}$ c. $\dfrac{32}{4}$ d. $\dfrac{18}{3}$ e. $\dfrac{18}{9}$

Part 3

A person has a big bag of coins with lots of pennies, lots of nickels and lots of dimes.

a. What's the smallest number of cents the person could have with 4 coins?

b. What's the largest number of cents the person could have with 4 coins?

c. A person has 26 cents with 4 coins. What are the 4 coins?

d. The person had 35 cents with 4 coins. What are the 4 coins?

Part 4

A person has a big bag of coins. The only coins in the bag are nickels, dimes and quarters.

a. What's the smallest number of cents the person could have with 3 coins?

b. What's the largest number of cents the person could have with 3 coins?

c. The person has 55 cents with 3 coins. What are the 3 coins?

d. The person has 45 cents with 3 coins. What are the 3 coins?

e. The person has 35 cents with 3 coins. What are the 3 coins?

Independent Work

Part 5 For the problems with whole numbers, make the number family, write the column problem and figure out the answer. For the problems with fractions, write the addition or subtraction problem and the answer. Remember the unit name.

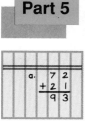

a. A plant was $\frac{11}{3}$ yards tall. A gardener cuts $\frac{6}{3}$ yards off the top of the plant. How tall did the plant end up?

b. Joe's kitten weighed $\frac{13}{5}$ pounds. The kitten gained $\frac{3}{5}$ pound. How much did Joe's kitten end up weighing?

c. A truck started out with some hay on it. 538 pounds of hay were loaded on the truck. Now the truck has 1060 pounds of hay. How much hay was in the truck at first?

d. A rabbit ran 318 fewer feet than a tortoise. The rabbit ran 343 feet. How far did the tortoise run?

Part 6 Copy the problems you can work and write the answers.

a. $\frac{11}{3} - \frac{11}{2} =$

b. $\frac{15}{5} + \frac{5}{8} =$

c. $\frac{3}{8} - \frac{3}{8} =$

d. $\frac{7}{4} + \frac{9}{4} =$

e. $\frac{17}{17} - \frac{3}{17} =$

f. $\frac{7}{6} - \frac{6}{7} =$

Part 7 Copy each equation and complete it.

a. $47 - \blacksquare = 8 \times 0$

b. $3 + 5 + 2 = 11 - \blacksquare$

c. $16 \blacksquare 4 = 5 \times 4$

d. $7 \times 5 = 40 \blacksquare 5$

Part 8   Copy the table. Use the facts to put numbers in the table. Then figure out the missing numbers and answer the questions.

This table is supposed to show how much Rose and Amy spent on Saturday and Sunday.

	Saturday	Sunday	Total for both days
Rose			
Amy			
Total for both girls			

Questions

a. How much more money did the girls spend on Sunday than on Saturday?

b. How much less did Rose spend on Saturday than on Sunday?

c. How much more money did Amy spend than Rose spent on both days?

Fact 1: Amy spent $1.87 on Saturday.

Fact 2: Amy spent $1.44 on Sunday.

Fact 3: On Saturday, Rose spent $1.12 less than Amy.

Fact 4: Rose spent a total of $3.05 on both days.

Part 9   Figure out the route for each problem and write the number of miles.

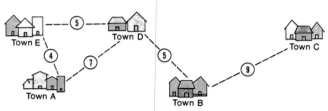

There is more than one way to go from town E to town B.

a. Figure out the route that is the longest and write the number of miles for that route.

b. Figure out the route that is the shortest and write the number of miles for that route.

Lesson 112

Part 1

a. There were 7 windows in each room. There were 19 rooms. How many windows were there?

b. They put 31 bugs in each carton. They had 5 cartons. How many bugs did they have in all?

c. There are 4 desks in each room. There are 32 desks in all. How many rooms are there?

d. There were 9 spots on each bug. There were 63 spots in all. How many bugs were there?

Part 2 Paired Practice

a. 8 x 3 = ■   b. 3 x 3 = ■   c. 2 x 3 = ■   d. 3 x 6 = ■

e. 10 x 3 = ■   f. 3 x 7 = ■   g. 7 x 3 = ■   h. 9 x 3 = ■

i. 3 x 8 = ■   j. 5 x 3 = ■   k. 3 x 4 = ■   l. 3 x 10 = ■

m. 6 x 3 = ■   n. 3 x 5= ■   o. 4 x 3 = ■   p. 3 x 9 = ■

q. 3 x 3 = ■

This stove costs $10⁰⁰ less than the other one.

$450

- This is the first part of a table problem that involves time.

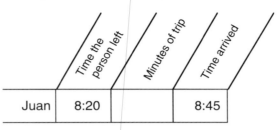

	Time the person left	Minutes of trip	Time arrived
Juan	8:20		8:45

- The first column shows the time Juan left on a trip. Juan left at 8:20.

- The second column is supposed to show the number of minutes the trip took.

- The last column shows the time Juan arrived. He arrived at 8:45.

- Remember, each row in a table works just like a number family. The first two numbers are the small numbers. The last number is the big number.

- This is the number family for Juan: $\underrightarrow{8:20 \qquad \square} 8:45$

Part 4   For each fraction write the division problem and the answer.

a. $\dfrac{15}{3}$ b. $\dfrac{54}{6}$ c. $\dfrac{32}{8}$ d. $\dfrac{12}{2}$ e. $\dfrac{70}{10}$ f. $\dfrac{40}{5}$

Part 5   Make a number family for each problem. Figure out the answers. Remember the unit name.

a. A jar contains red marbles and blue marbles. There are 177 blue marbles. If there are 263 marbles in the jar, how many red marbles are there?

b. Jim weighs 48 pounds more than Sue. Jim weighs 215 pounds. How much does Sue weigh?

c. A boat travels 168 miles per hour slower than a plane. The boat travels 46 miles per hour. How fast does the plane travel?

d. Gold fish and angel fish live in a pond. There are 424 fish in the pond. If there are 77 angel fish, how many gold fish are there?

Part 6   Write what X equals and what Y equals for each letter.

A. (X = ■, Y = ■)

B. (X = ■, Y = ■)

C. (X = ■, Y = ■)

D. (X = ■, Y = ■)

Copy each problem. Complete the sign.

a. $\dfrac{7}{7}$ ___ $\dfrac{1}{7}$ b. $\dfrac{8}{5}$ ___ $\dfrac{6}{5}$ c. $\dfrac{1}{9}$ ___ $\dfrac{2}{9}$ d. $\dfrac{5}{5}$ ___ 1

Part 8 Rewrite each problem as a column problem and solve it.

a. 259 x 7 =   b. 36 x 4 =   c. 506 x 3 =

```
a.  7 2
   +2 1
    9 3
```

Part 9 Rewrite each problem as a column problem and solve it.

a. 506 + 3 =   c. 36 – 28 =

b. 259 – 187 =   d. 830 + 92 =

```
a.  7 2
   +2 1
    9 3
```

Part 10 Write the answer to each problem.

a. 7   b. 11   c. $9\overline{\smash{)}9}$ d. 0   e. 18   f. 0   g. 4
   +0   – 4   x 7   – 0   + 35   x 4

Lesson 113

Part 1

- These are problems that a lot of high school students have trouble with. But the problems are just like the addition and subtraction problems you've worked with number families.

- Problem A: If <u>Tom has 164 fewer marbles than Mary</u>, and if <u>Mary has 396 marbles</u>, <u>how many marbles does Tom have</u>?

- The underlined parts show that there are three parts to the problem.

- The first underlined part says: <u>Tom has 164 fewer marbles than Mary</u>. That part tells how to make the number family.

- The next underlined part says: <u>Mary has 396 marbles</u>. That part tells about a number that belongs in the family.

- The last underlined part says: <u>How many marbles does Tom have</u>? That part asks the question. You make a number family for this problem and figure out the answer.

a. If <u>Tom has 164 fewer marbles than Mary</u>, and if <u>Mary has 396 marbles</u>, <u>how many marbles does Tom have</u>?

b. If <u>the school is 13 miles closer than the farm</u>, and if <u>the farm is 29 miles away</u>, <u>how far away is the school</u>?

c. If <u>the mule weighed 205 pounds less than the cow</u>, and if <u>the cow weighed 965 pounds</u>, <u>how much did the mule weigh</u>?

d. If the train went 34 miles farther than the bus went, and if the bus went 203 miles, how far did the train go?

a. There were 3 camels on each hill. There were 21 camels in all. How many hills were there?

b. Each tank held 4 gallons. There were 12 tanks. How many gallons were there?

c. There were 9 bugs in each tree. There were 72 bugs in all. How many trees were there?

d. They put 29 chairs in each room. There were 5 rooms. How many chairs were there?

Independent Work

Part 3 Copy each equation and complete it.

a. $15 - \boxed{} = 10 + 0$

b. $1 \times 0 = 8 - \boxed{}$

c. $11 + 20 = 1 \boxed{} 30$

d. $19 + 1 + \boxed{} = 5 \times 4$

Part 4 Write the fraction for each problem.

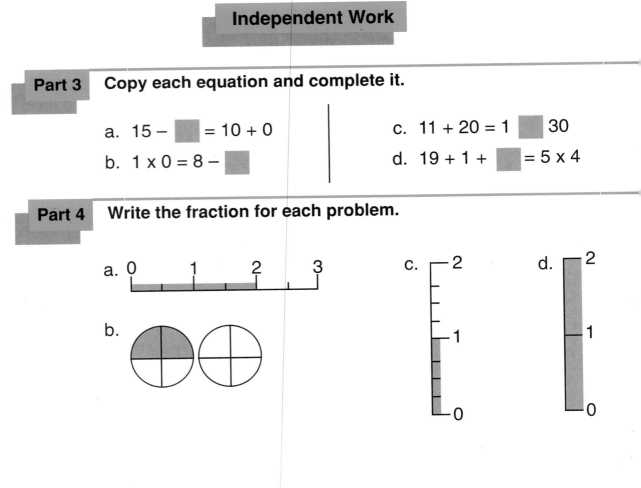

For each item, write the addition or subtraction problem and figure out the answer.

a. Char's hair was $\frac{7}{3}$ feet long. Her hair grew $\frac{1}{3}$ foot. How long did Char's hair end up being?

b. A bucket contained $\frac{9}{4}$ bushels of corn. A horse ate $\frac{5}{4}$ bushels of corn out of the bucket. How much corn was left in the bucket?

c. There were $\frac{11}{5}$ tons of dirt in the pile. A truck dumped $\frac{32}{5}$ more tons of dirt on the pile. How much dirt ended up in the pile?

Part 6 Write the answer to each problem.

A person has a big bag of coins with pennies, nickels, and dimes.

a. What's the smallest number of cents the person could have with 3 coins?

b. What's the largest number of cents the person could have with 3 coins?

c. A person has 3 coins worth 12 cents. What are the coins?

d. The person had 3 coins worth 20 cents. What are the coins?

Part 7 Rewrite each problem as a column problem and solve it.

a. 43 x 3 =   b. 70 x 9 =   c. 806 x 5 =

Lesson 114

Part 1

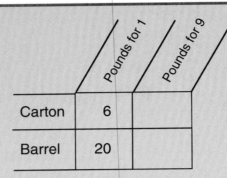

	Pounds for 1	Pounds for 9
Carton	6	
Barrel	20	

- You're going to use the table to answer questions like this one: **How much do 9 cartons and 1 barrel weigh?**

- That's an addition problem. You know the weight of 9 cartons and the weight of 1 barrel. You add those amounts to figure out the answer:

$$\begin{array}{r} 54 \longleftarrow \text{9 cartons} \\ + 20 \longleftarrow \text{1 barrel} \\ \hline 74 \end{array}$$

- 9 cartons and 1 barrel weigh 74 pounds.

a. How much do 1 carton and 1 barrel weigh?

b. How much more do 9 barrels weigh than 9 cartons?

c. How much do 1 carton and 9 barrels weigh?

Part 2 **For each problem, make the number family and figure out the answer.**

a. If a train travels 161 miles farther than the car, and if the car travels 310 miles, how far does the train travel?

b. If a woman spends 32 dollars more than her husband spends, and if the woman spends 234 dollars, how much does her husband spend?

c. If a ship weighs 12 tons more than a truck, and if the truck weighs 21 tons, how much does the ship weigh?

d. If the lodge is 17 miles closer than the mountain, and if the lodge is 79 miles away, how far away is the mountain?

e. If the tank holds 350 more gallons than the barrel, and if the tank holds 536 gallons, how many gallons does the barrel hold?

Part 3 **For each problem, write the number family and figure out the answer.**

a. 5 inches of rain fell each day. 45 inches of rain fell in all. On how many days did it rain?

b. There were 25 birds in each tree. There were 9 trees. How many birds were there in all?

c. Each basket held 30 chicks. There were 3 baskets. How many chicks were there in all?

d. They had 7 chicks in each basket. They had 35 chicks in all. How many baskets did they have?

Independent Work

Part 4 **Copy each problem and figure out the answer.**

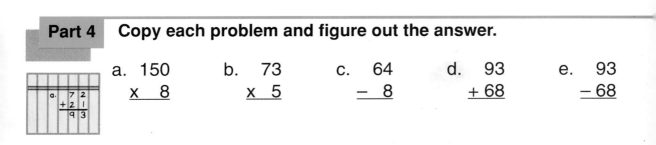

a. 150   b. 73   c. 64   d. 93   e. 93
   x 8   x 5   − 8   + 68   − 68

Part 5   Write the answer to each problem.

a.   9
    − 4

b.   15
    + 0

c.   0
    x 6

d.   13
    − 6

e.   4 $\overline{)32}$

f.   1
    x 9

g.   5 $\overline{)35}$

h.   3
    x 6

i.   10
    − 0

Part 6   Copy each problem. Complete the sign.

a. $\frac{8}{3}$ __ $\frac{7}{3}$

b. $\frac{2}{6}$ __ $\frac{3}{6}$

c.   1 __ $\frac{9}{8}$

d. $\frac{5}{5}$ __ $\frac{8}{8}$

Part 7   Figure out the route for each problem and write the number of miles.

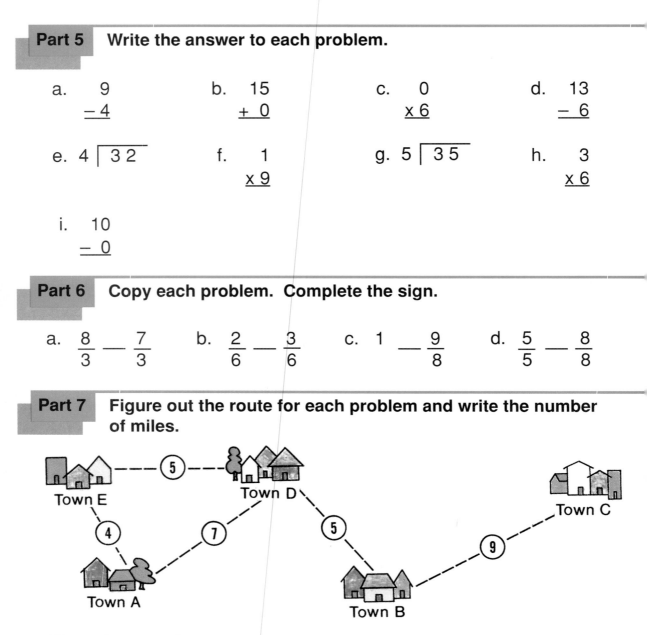

There is more than one way to go from town A to town C.

a. Figure out the route that is the longest and write the number of miles for that route.

b. Figure out the route that is the shortest and write the number of miles for that route.

Lesson 115

Part 1

	Cents for 1	Cents for 9
Nickels		
Dimes		
Quarters		

a. How many cents are 9 dimes and 1 quarter worth?

b. How much more are 9 quarters worth than 9 nickels are worth?

c. How many cents are 9 nickels, 1 dime and 1 quarter worth?

d. How much more are 9 nickels worth than 1 quarter is worth?

Part 2 Make a number family for each word problem. Then write the number problem and the answer.

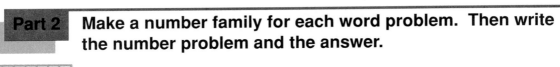

a. If <u>a dog has 56 more fleas than a cat</u>, and if <u>the dog has 113 fleas</u>, <u>how many fleas does the cat have</u>?

b. If a train weighs 36 tons less than a ship, and if the train weighs 113 tons, how much does the ship weigh?

c. If a girl is 24 years younger than her mother, and if her mother is 31 years old, how old is the girl?

d. If a mother is 32 years older than her son, and if the mother is 40 years old, how old is the son?

- This fraction is 14-sevenths: $\dfrac{14}{7}$
- You can write the fraction as a division problem: $7\overline{)14}$
- You can also write the division problem another way, using this sign: ÷
- The sign looks like a fraction. The dot on top stands for the top number of the fraction. The dot on the bottom stands for the bottom number.
- Here's $7\overline{)14}$ written with the divided-by sign: $14 \div 7$.

Independent Work

Part 4 For these problems you multiply, add, or subtract. Copy each probem and figure out the answer.

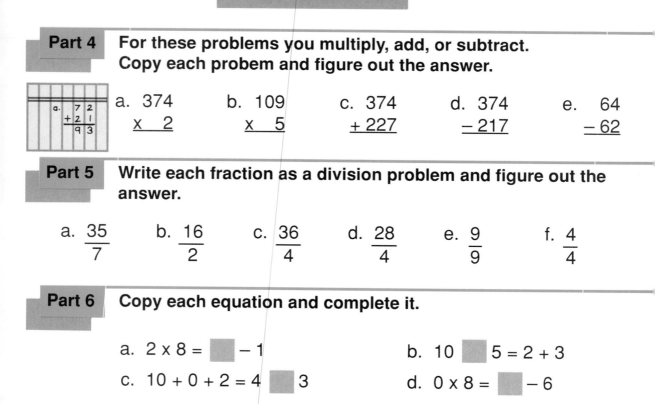

	a. 374	b. 109	c. 374	d. 374	e. 64
	× 2	× 5	+ 227	− 217	− 62

Part 5 Write each fraction as a division problem and figure out the answer.

a. $\dfrac{35}{7}$ b. $\dfrac{16}{2}$ c. $\dfrac{36}{4}$ d. $\dfrac{28}{4}$ e. $\dfrac{9}{9}$ f. $\dfrac{4}{4}$

Part 6 Copy each equation and complete it.

a. $2 \times 8 = \boxed{} - 1$

b. $10 \boxed{} 5 = 2 + 3$

c. $10 + 0 + 2 = 4 \boxed{} 3$

d. $0 \times 8 = \boxed{} - 6$

Part 7 — Write the answer to each question.

> A person has a bag of pennies, nickels, and dimes.

a. What's the smallest number of cents the person could have with 4 coins?

b. What's the largest number of cents the person could have with 4 coins?

c. A person has 30 cents with 4 coins. What are the coins?

Part 8

a. Valerie buys juice, milk and cereal. **About** how much does she spend?

b. A man buys bananas, meat and milk. **Exactly** how much does he spend?

c. A boy buys bananas, juice and meat. **About** how much does he spend?

d. A boy buys bananas, juice and meat. **Exactly** how much does he spend?

Part 9 — Write the fraction for each problem.

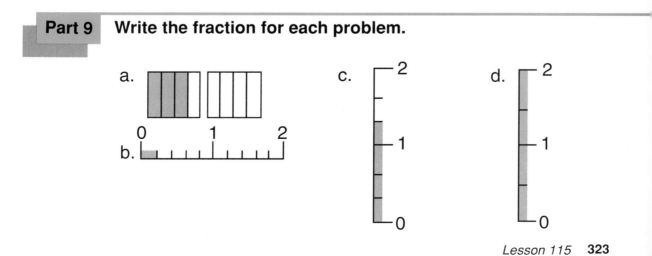

Lesson 116

Part 1 **Make a number family for each word problem. Then write the number problem and the answer.**

a. If the barn is 300 feet farther away than the fence, and if the fence is 451 feet away, how far away is the barn?

b. If the house is 27 feet closer than the barn, and if the barn is 96 feet away, how far away is the house?

c. If Jan is 25 years younger than her mother, and if Jan is 14 years old, how old is her mother?

d. If Sam is 29 years older than Jim, and if Sam is 50 years old, how old is Jim?

Part 2

a. 3 ⟌ 2 7   b. 3 ⟌ 1 8   c. 3 ⟌ 2 4   d. 3 ⟌ 2 1

e. 3 ⟌ 9   f. 3 ⟌ 1 5   g. 3 ⟌ 1 2   h. 3 ⟌ 2 4

i. 3 ⟌ 1 8   j. 3 ⟌ 2 7   k. 3 ⟌ 2 1

Part 3

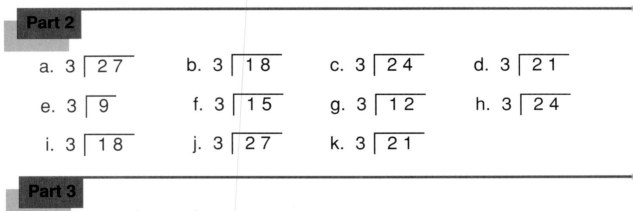

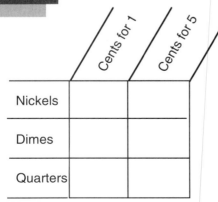

a. How much less are 5 nickels worth than 5 quarters are worth?

b. How many cents are 1 nickel, 5 dimes and 5 quarters worth?

c. How many cents are 5 nickels, 5 dimes and 1 quarter worth?

d. How much less are 5 dimes worth than 5 quarters are worth?

- Some problems that compare have fractions.

- You work them just like any other problem that compares. You put the fractions in a number family. Then you add or subtract to find the missing number.

- Here's a problem: **A hippo weighed $\frac{3}{4}$ ton less than a truck. The hippo weighed $\frac{9}{4}$ tons. How much did the truck weigh?**

- Here's the number family for the first sentence:

$$\frac{3}{4} \xrightarrow{\quad H \quad} T$$

- The truck weighed more, $\frac{3}{4}$ ton more. So **truck** is the big number. **Hippo** and $\frac{3}{4}$ are the small numbers.

- Then you put in the number you know. The hippo weighed $\frac{9}{4}$ tons.

$$\frac{3}{4} \xrightarrow{\quad \overset{9/4}{\cancel{H}} \quad} T$$

- Now you write the number problem and figure out the answer.

$$\frac{3}{4} + \frac{9}{4} = \frac{12}{4}$$

- The truck weighed $\frac{12}{4}$ tons.

Make a number family for each word problem. Then write the number problem and the answer.

a. John ran $\frac{5}{3}$ fewer miles than Meg ran. John ran $\frac{14}{3}$ miles. How far did Meg run?

b. A bottle weighs $\frac{2}{5}$ pound more than a can weighs. The can weighs $\frac{7}{5}$ pounds. How many pounds does the bottle weigh?

Part 5 Copy each problem. Complete the sign.

a. $\dfrac{1}{3}$ ___ 1

b. $\dfrac{4}{7}$ ___ $\dfrac{3}{7}$

c. 1 ___ $\dfrac{8}{8}$

d. $\dfrac{4}{4}$ ___ $\dfrac{8}{4}$

Part 6 Copy each problem and figure out the answer.

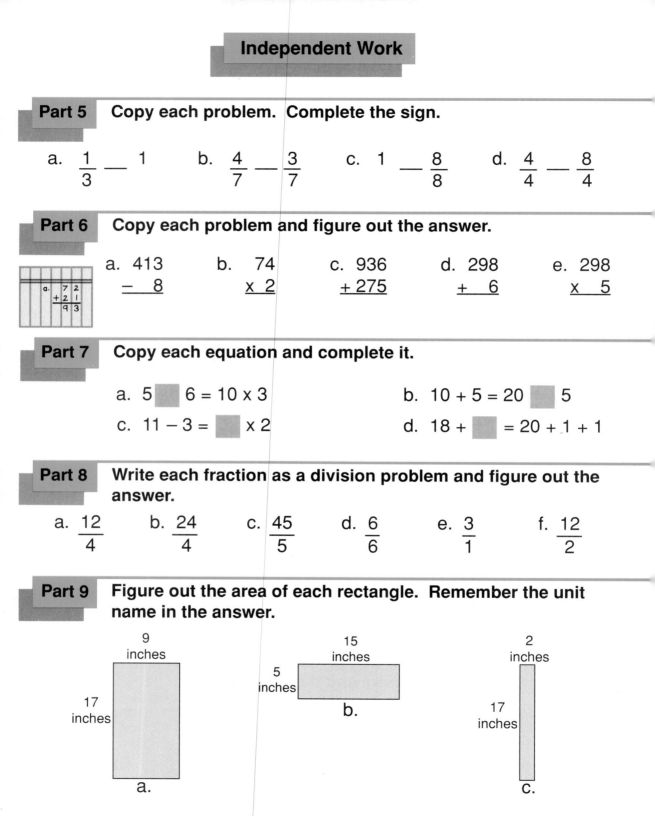

a. 413
 − 8

b.   74
 x 2

c. 936
 + 275

d. 298
 +   6

e. 298
 x   5

Part 7 Copy each equation and complete it.

a. 5 ▨ 6 = 10 x 3

b. 10 + 5 = 20 ▨ 5

c. 11 − 3 = ▨ x 2

d. 18 + ▨ = 20 + 1 + 1

Part 8 Write each fraction as a division problem and figure out the answer.

a. $\dfrac{12}{4}$

b. $\dfrac{24}{4}$

c. $\dfrac{45}{5}$

d. $\dfrac{6}{6}$

e. $\dfrac{3}{1}$

f. $\dfrac{12}{2}$

Part 9 Figure out the area of each rectangle. Remember the unit name in the answer.

9 inches

17 inches

a.

15 inches

5 inches

b.

2 inches

17 inches

c.

Lesson 117

Part 1

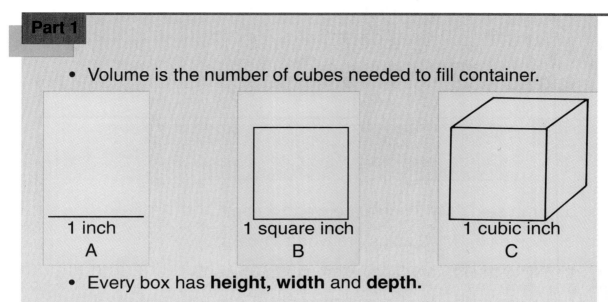

- Volume is the number of cubes needed to fill container.

1 inch	1 square inch	1 cubic inch
A	B	C

- Every box has **height, width** and **depth.**
- The **depth** is how far back it goes.
- To find the volume of a box, you multiply **height** times **width** times **depth.**
- The units in the answer are cubic units, not square units.

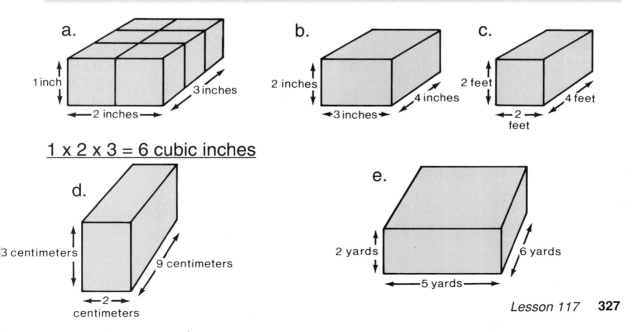

a. 1 inch, 2 inches, 3 inches

1 x 2 x 3 = 6 cubic inches

b. 2 inches, 3 inches, 4 inches

c. 2 feet, 2 feet, 4 feet

d. 3 centimeters, 2 centimeters, 9 centimeters

e. 2 yards, 5 yards, 6 yards

Part 2

a. 3⟌9 b. 3⟌24 c. 3⟌6 d. 3⟌21

e. 3⟌12 f. 3⟌27 g. 3⟌24 h. 3⟌15

i. 3⟌21 j. 3⟌18 k. 3⟌27 l. 3⟌18

Part 3

A bag has pennies, nickels and dimes.
Different people reach in the bag and pull out 3 coins.

a. Fran said, "My 3 coins are worth 2 cents."

b. Ron said, "My 3 coins are worth 11 cents."

c. Ginger said, "My 3 coins are worth 35 cents."

d. Donna said, "My 3 coins are worth 7 cents."

e. Sally said, "My 3 coins are worth 21 cents."

Independent Work

Part 4 Copy each problem and write the answer.

a. $14 \div 7 =$ ■ b. $14 - 7 =$ ■ c. $8 + 6 =$ ■ d. $4 \times 0 =$ ■

e. $3 \times 8 =$ ■ f. $11 - 6 =$ ■ g. $7 \times 1 =$ ■ h. $45 \div 5 =$ ■

i. $4 + 0 =$ ■ j. $0 \times 5 =$ ■

Part 5 Copy each problem and figure out the answer.

a. 470   b. 45   c. 483   d. 403   e. 304
 − 283   x 8   + 270   x   9   −   8

```
a.  7 2
  + 2 1
    9 3
```

Write the addition problems and the answers with a dollar sign.

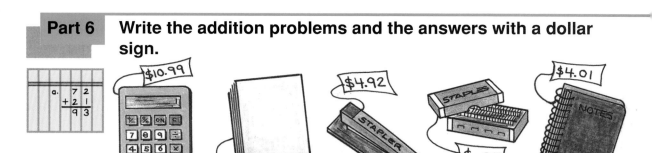

a. A boy bought a notebook, paper and a calculator. **Exactly** how much did he spend?

b. A girl bought a notebook, paper and a stapler. **About** how much did she spend?

c. Susie bought a calculator, a stapler and staples. **Exactly** how much did she spend?

d. Susie bought paper and a notebook. **About** how much did she spend?

Part 7   For each problem, write the number family and figure out the answer.

a. Each quarter is worth 25 cents. Jill has 7 quarters. How many cents in quarters does Jill have?

b. There are 36 cookies in every box. A store has 5 boxes of cookies. How many cookies are there?

Part 8   Copy the problems you can work. Then write the answers.

a. $\dfrac{11}{3} + \dfrac{8}{3} =$

b. $\dfrac{6}{6} + \dfrac{6}{6} =$

c. $\dfrac{2}{4} - \dfrac{2}{2} =$

d. $\dfrac{9}{7} - \dfrac{7}{7} =$

Figure out the route for each problem and write the number of miles.

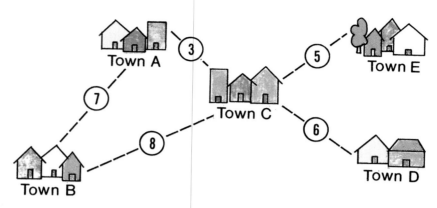

a. Figure out the longest route from town B to town D.

b. Figure out the shortest route from town B to town D.

c. Figure out the longest route from town A to town E.

d. Figure out the shortest route from town A to town E.

MAMMOTH MOVIES
STUDIO CAFETERIA

Part 1

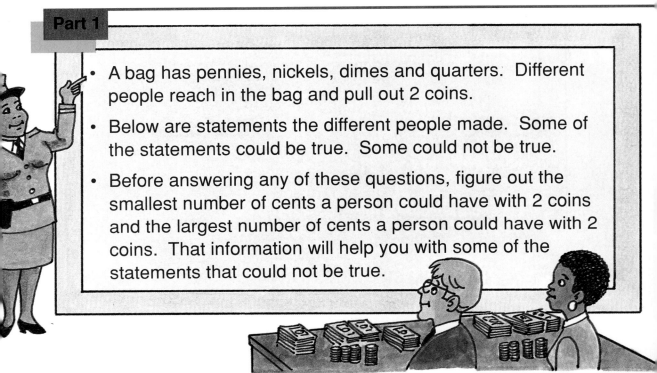

- A bag has pennies, nickels, dimes and quarters. Different people reach in the bag and pull out 2 coins.

- Below are statements the different people made. Some of the statements could be true. Some could not be true.

- Before answering any of these questions, figure out the smallest number of cents a person could have with 2 coins and the largest number of cents a person could have with 2 coins. That information will help you with some of the statements that could not be true.

a. Ron said, "My 2 coins are worth 20 cents."

b. Fran said, "My 2 coins are worth 5 cents."

c. Ginger said, "My 2 coins are worth 35 cents."

d. Donna said, "My 2 coins are worth 70 cents."

e. Sally said, "My 2 coins are worth 50 cents."

Part 2   Work these problems on your calculator.

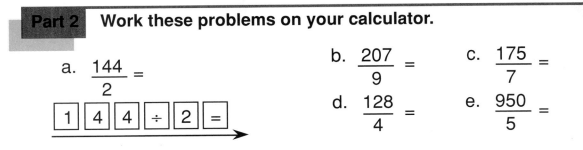

a. $\dfrac{144}{2} =$

$\boxed{1}\ \boxed{4}\ \boxed{4}\ \boxed{\div}\ \boxed{2}\ \boxed{=}$

b. $\dfrac{207}{9} =$

c. $\dfrac{175}{7} =$

d. $\dfrac{128}{4} =$

e. $\dfrac{950}{5} =$

Write the multiplication problem and the answer for each box.

a.

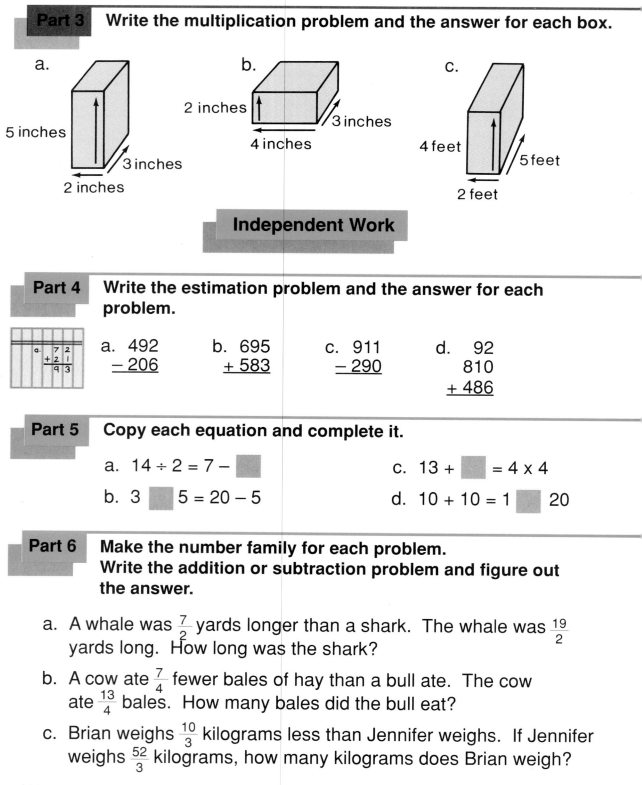

5 inches

3 inches

2 inches

b.

2 inches

3 inches

4 inches

c.

4 feet

5 feet

2 feet

Independent Work

Part 4 Write the estimation problem and the answer for each problem.

a. 492
  − 206

b. 695
  + 583

c. 911
  − 290

d.   92
  810
  + 486

Part 5 Copy each equation and complete it.

a. 14 ÷ 2 = 7 − ▨

b. 3 ▨ 5 = 20 − 5

c. 13 + ▨ = 4 x 4

d. 10 + 10 = 1 ▨ 20

Part 6 Make the number family for each problem.
Write the addition or subtraction problem and figure out the answer.

a. A whale was $\frac{7}{2}$ yards longer than a shark. The whale was $\frac{19}{2}$ yards long. How long was the shark?

b. A cow ate $\frac{7}{4}$ fewer bales of hay than a bull ate. The cow ate $\frac{13}{4}$ bales. How many bales did the bull eat?

c. Brian weighs $\frac{10}{3}$ kilograms less than Jennifer weighs. If Jennifer weighs $\frac{52}{3}$ kilograms, how many kilograms does Brian weigh?

Part 7 Copy the table. Use the facts to put numbers in the table. Then answer the questions.

This table is supposed to show the lunches and dinners Chin's Cafe and Ruby's Restaurant served.

	Chin's Cafe	Ruby's Restaurant	Total for both places
Lunches			
Dinners			421
Total meals			

Questions

a. How many more meals did Ruby's Restaurant serve than Chin's Cafe served?

b. How many more lunches than dinners were served?

c. How many more lunches were served at Chin's Cafe than at Ruby's Restaurant?

Fact 1: Chin's Cafe served 442 meals.

Fact 2: Ruby's Restaurant served 237 dinners.

Fact 3: Chin's Cafe served 53 fewer dinners than Ruby's Restaurant served.

Fact 4: Ruby's Restaurant served 8 more dinners than lunches.

Part 8 Write the fraction for each problem.

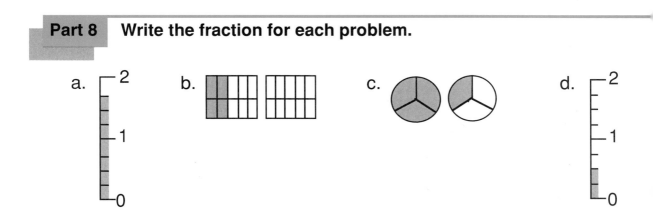

a.

b.

c.

d.

Part 9 Copy each problem and figure out the answer.

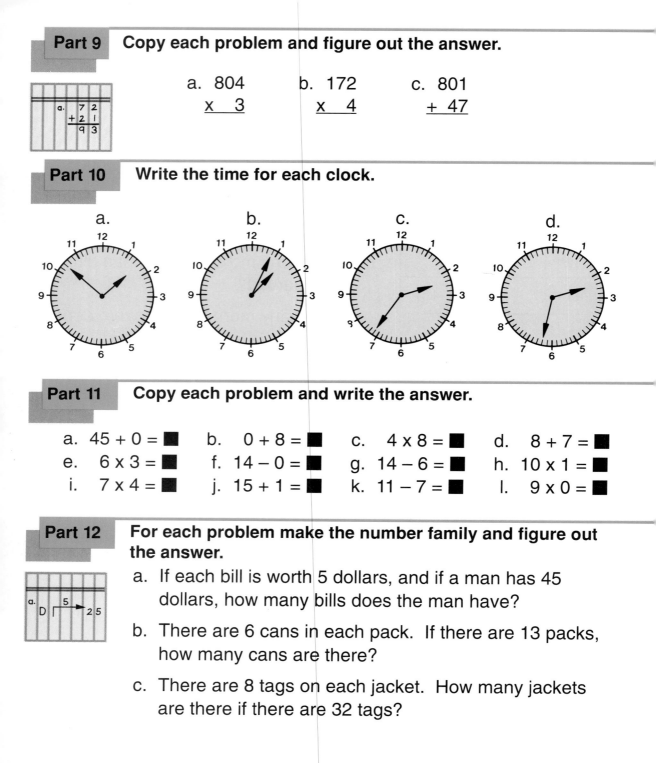

	a.	804	b.	172	c.	801
		x 3		x 4		+ 47

Part 10 Write the time for each clock.

a.

b.

c.

d.

Part 11 Copy each problem and write the answer.

a. 45 + 0 = ■   b. 0 + 8 = ■   c. 4 x 8 = ■   d. 8 + 7 = ■

e. 6 x 3 = ■   f. 14 − 0 = ■   g. 14 − 6 = ■   h. 10 x 1 = ■

i. 7 x 4 = ■   j. 15 + 1 = ■   k. 11 − 7 = ■   l. 9 x 0 = ■

Part 12 For each problem make the number family and figure out the answer.

a. If each bill is worth 5 dollars, and if a man has 45 dollars, how many bills does the man have?

b. There are 6 cans in each pack. If there are 13 packs, how many cans are there?

c. There are 8 tags on each jacket. How many jackets are there if there are 32 tags?

For each fraction write the division problem and the answer.

a. $\dfrac{32}{8}$ b. $\dfrac{70}{7}$ c. $\dfrac{15}{3}$ d. $\dfrac{45}{5}$ e. $\dfrac{40}{8}$ f. $\dfrac{27}{9}$

g. $\dfrac{81}{9}$ h. $\dfrac{24}{4}$ i. $\dfrac{16}{8}$ j. $\dfrac{16}{4}$ k. $\dfrac{8}{8}$

Do the independent work for lesson 118 in your workbook.

Lesson 119

Part 1

- You can read fractions as division problems.
- That means you can work with fractions using a calculator.
- You'll use the divided-by sign on your calculator.

a. $\dfrac{144}{9} =$ b. $\dfrac{460}{10} =$ c. $\dfrac{237}{3} =$ d. $\dfrac{144}{6} =$

Part 2

a. $5\overline{)38}$ R   b. $9\overline{)38}$ R   c. $4\overline{)9}$ R   d. $4\overline{)19}$ R   e. $4\overline{)29}$ R

Part 3

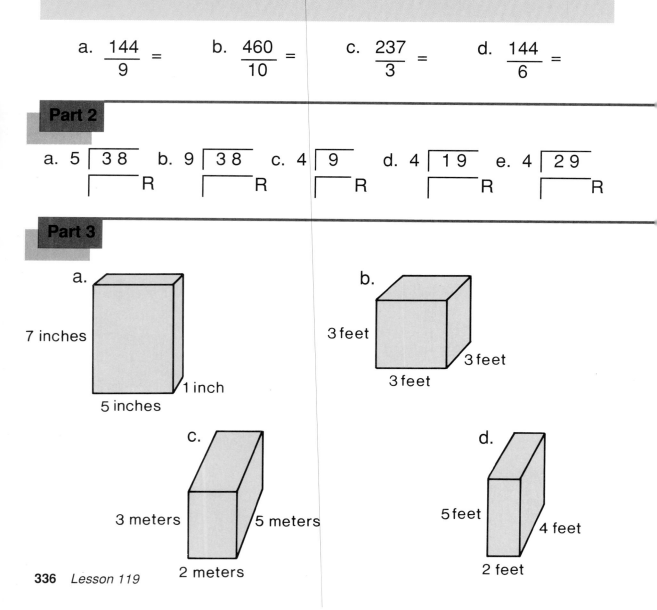

a. 7 inches, 5 inches, 1 inch

b. 3 feet, 3 feet, 3 feet

c. 3 meters, 5 meters, 2 meters

d. 5 feet, 4 feet, 2 feet

Paired Practice

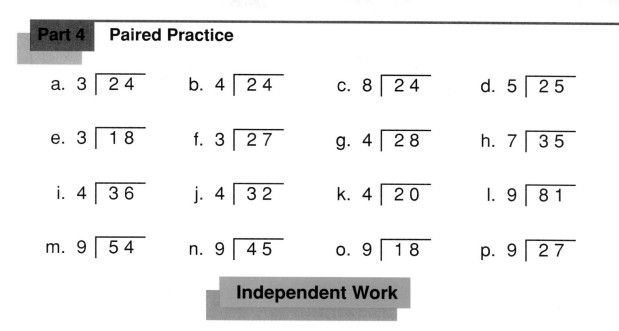

a. 3 ⟌ 2 4   b. 4 ⟌ 2 4   c. 8 ⟌ 2 4   d. 5 ⟌ 2 5

e. 3 ⟌ 1 8   f. 3 ⟌ 2 7   g. 4 ⟌ 2 8   h. 7 ⟌ 3 5

i. 4 ⟌ 3 6   j. 4 ⟌ 3 2   k. 4 ⟌ 2 0   l. 9 ⟌ 8 1

m. 9 ⟌ 5 4   n. 9 ⟌ 4 5   o. 9 ⟌ 1 8   p. 9 ⟌ 2 7

Independent Work

Part 5 **Make a number family for each word problem. Then write the number problem and the answer.**

a. The cow weighed 846 pounds more than the pig weighed. If the pig weighed 588 pounds, how much did the cow weigh?

b. Joe had $6.78 less than Sue had. How much money did Joe have if Sue had $7.34?

c. Ted worked 173 hours longer than Robert worked. How long did Ted work if Robert worked 602 hours?

Part 6 **Make a number family for each word probem. Then write the number problem and the answer.**

a. Each scissor weighs 9 ounces. How much do 15 scissors weigh?

b. Each bucket holds 4 cans. How many cans do 50 buckets hold?

Copy each problem. Complete the sign.

a. $\dfrac{4}{4}$ ___ $\dfrac{3}{4}$ b. $\dfrac{6}{5}$ ___ $\dfrac{4}{5}$ c. $\dfrac{8}{7}$ ___ $\dfrac{7}{7}$ d. $\dfrac{5}{5}$ ___ $\dfrac{9}{9}$

Part 8 Write each problem as a column problem and figure out the answer.

a.
```
   7 2
 + 2 1
 -----
   9 3
```

a. $573 - 8 =$ b. $925 \times 6 =$ c. $728 - 365 =$ d. $106 + 5 =$

Part 9 Write what X equals and what Y equals for each letter.

A. (X = ■, Y = ■)

B. (X = ■, Y = ■)

C. (X = ■, Y = ■)

D. (X = ■, Y = ■)

Part 10 If the statement could be true, write the number of **pennies, nickels** or **dimes** that person has. If the statement could be false, write **false.**

> A person has a bag of pennies, nickels and dimes. Each person has 3 coins.

a. Kay said, "My coins are worth 3 cents."

b. Ben said, "My coins are worth 7 cents."

c. Dan said, "My coins are worth 40 cents."

d. Sally said, "My coins are worth 15 cents."

e. Joan said, "My coins are worth 16 cents."

Part 11   **Copy each equation and complete it.**

a. $5 \div 1 = 10 \,\blacksquare\, 5$ b. $16 - \blacksquare = 9 + 1$

c. $18 + \blacksquare + 1 = 4 \times 5$ d. $40 - 40 = 39 \,\blacksquare\, 0$

Part 12

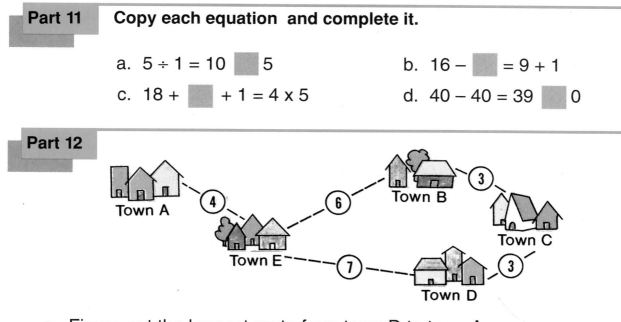

a. Figure out the longest route from town D to town A.

b. Figure out the shortest route from town D to town A.

Part 13   **Copy each problem you can work and write the answer.**

a. $\dfrac{4}{5} - \dfrac{4}{6} =$ b. $\dfrac{8}{10} + \dfrac{10}{10} =$ c. $\dfrac{10}{2} - \dfrac{5}{2} =$

d. $\dfrac{7}{4} + \dfrac{7}{7} =$ e. $\dfrac{9}{4} + \dfrac{8}{1} =$ f. $\dfrac{6}{3} - \dfrac{2}{3} =$

Part 14   **Write an answer for each problem.**

a. $35 \div 7 = \blacksquare$ b. $15 + 8 = \blacksquare$ c. $3 \times 0 = \blacksquare$ d. $12 - 5 = \blacksquare$

e. $7 + 0 = \blacksquare$ f. $7 \times 4 = \blacksquare$ g. $32 \div 4 = \blacksquare$ h. $32 - 0 = \blacksquare$

i. $1 \times 32 = \blacksquare$ j. $0 \times 32 = \blacksquare$ k. $14 - 8 = \blacksquare$ l. $6 + 1 = \blacksquare$

m. $9 \div 1 = \blacksquare$ n. $36 \div 9 = \blacksquare$ o. $6 + 1 = \blacksquare$

Lesson 120

Part 1

a.

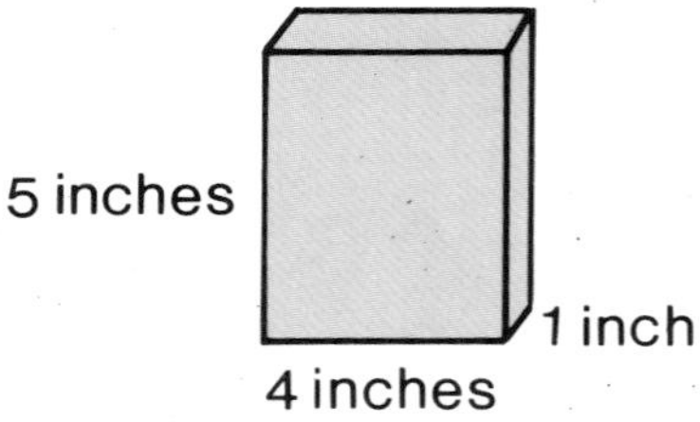

5 inches

1 inch

4 inches

b.

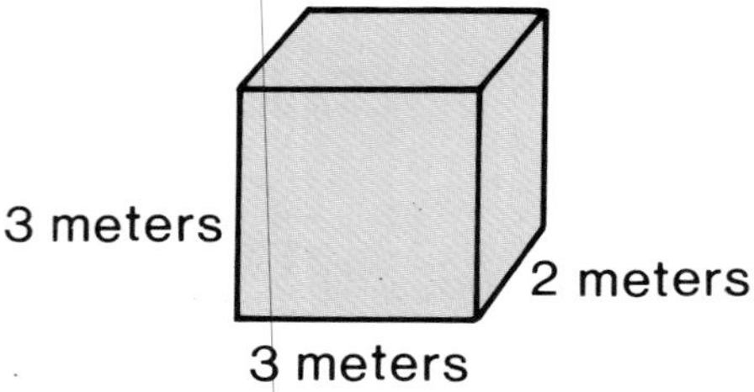

3 meters

2 meters

3 meters

c.

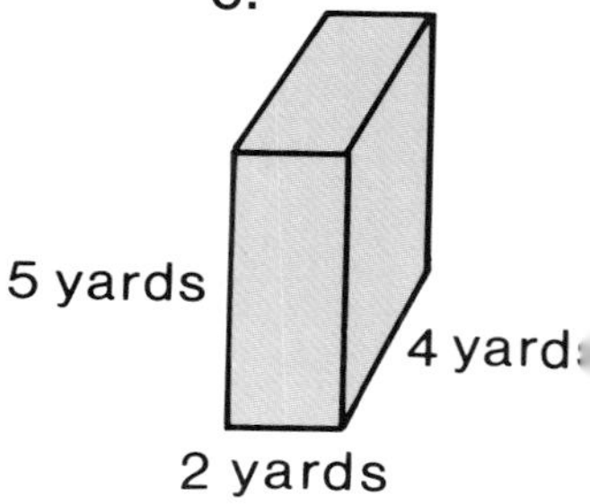

5 yards

4 yards

2 yards

Part 2   Work these problems on your calculator.

a. 169 ÷ 13

b. 21 x 7

c. 76 + 21

d. 13 x 6

e. 364 ÷ 13

f. 364 − 260

Part 3   Copy and work each problem.

a. 9 ⟌ 5 9   R

b. 9 ⟌ 7 9   R

c. 9 ⟌ 1 2   R

d. 4 ⟌ 2 6   R

e. 4 ⟌ 3 0   R

Part 4   Paired Practice

a. 9 ⟌ 7 2

b. 4 ⟌ 2 8

c. 4 ⟌ 1 6

d. 4 ⟌ 8

e. 10 ⟌ 4 0

f. 7 ⟌ 2 1

g. 4 ⟌ 2 4

h. 6 ⟌ 3 0

i. 9 ⟌ 2 7

j. 9 ⟌ 3 6

k. 8 ⟌ 3 2

l. 3 ⟌ 2 7

m. 3 ⟌ 2 1

n. 3 ⟌ 2 4

o. 3 ⟌ 1 8

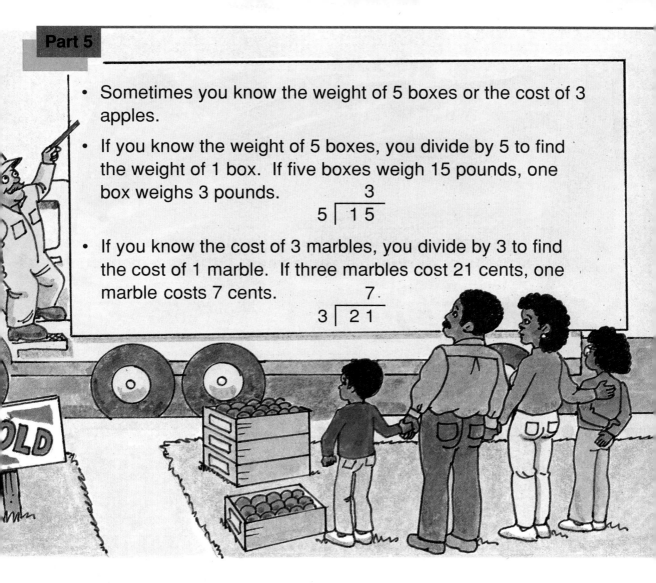

Part 5

- Sometimes you know the weight of 5 boxes or the cost of 3 apples.

- If you know the weight of 5 boxes, you divide by 5 to find the weight of 1 box. If five boxes weigh 15 pounds, one box weighs 3 pounds.

$$5\overline{)15} \quad ^3$$

- If you know the cost of 3 marbles, you divide by 3 to find the cost of 1 marble. If three marbles cost 21 cents, one marble costs 7 cents.

$$3\overline{)21} \quad ^7$$

Copy and complete each table.

a.

	Pounds for 1	Pounds for 5
Phone		20
Book		10
Chicken		35

b.

	Pounds for 1	Pounds for 3
Saw		24
Cart		30
Basket		21

Part 6 Write each problem as a division problem and figure out the answer.

a. $\dfrac{36}{4}$
b. $2\overline{\smash{)}16}$
c. $4\overline{\smash{)}20}$
d. $\dfrac{18}{6}$

Part 7 If the statement could be true, write the number of **pennies,** **nickels** or **dimes** that person has. If the statement could be false, write **false.**

> A person has a bag of pennies, nickels and dimes.
> Each person has 3 coins.

a. Dave said, "My coins are worth 70 cents."

b. Jill said, "My coins are worth 15 cents."

c. Carl said, "My coins are worth 25 cents."

d. Jim said, "My coins are worth 45 cents."

e. Sue said, "My coins are worth 50 cents."

Part 8 Make a multiplication number family for each word problem. Then figure out the answer.

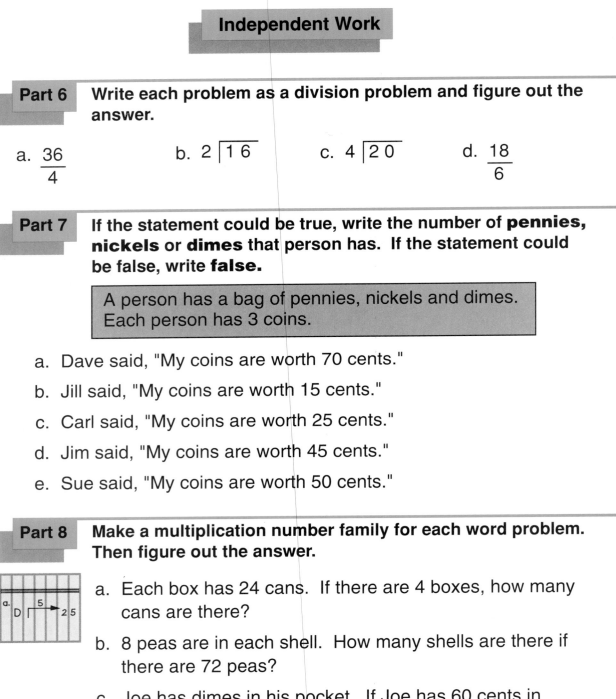

a. Each box has 24 cans. If there are 4 boxes, how many cans are there?

b. 8 peas are in each shell. How many shells are there if there are 72 peas?

c. Joe has dimes in his pocket. If Joe has 60 cents in dimes, how many dimes does he have in his pocket?

Part 9　Write the addition problems and answers with a dollar sign.

a. Sue wants to buy corn, fertilizer and a shovel. **Exactly** how much will she have to spend?

b. Carlos wants to buy a shovel and a rake. **About** how much will he have to spend?

c. Jan wants to buy tomato plants, fertilizer and a rake. **About** how much will she have to spend?

d. Tom wants to buy a rake, corn, a shovel and fertilizer. **Exactly** how much will he have to spend?

Part 10　Write each problem as a column problem and figure out the answer.

a. 408 x 2 =　　b. 742 − 667 =　　c. 408 + 2 =　　d. 542 x 7 =

Part 11　Make a number family for each problem. Then write the addition or subtraction problem and figure out the answer.

a. A restaurant had $\frac{15}{8}$ pies. Customers ate some pieces of pie. The restaurant ended up with $\frac{5}{8}$ of a pie. How many pies did the customers eat?

b. A rope was $\frac{13}{3}$ feet longer than a chain. The chain was $\frac{8}{3}$ feet long. How long was the rope?

c. A rock was $\frac{15}{5}$ pounds lighter than a stone. The stone weighed $\frac{19}{5}$ pounds. How much did the rock weigh?

d. Jill ran $\frac{2}{3}$ of a mile farther than Steve. If Steve ran $\frac{7}{3}$ miles, how far did Jill run?

Copy each equation and complete it.

a. $16 \div 2 = 10$ ▢ 2

b. $4 \times 3 \times 1 =$ ▢ $+ 2$

c. $18 -$ ▢ $= 18 \times 0$

d. 10 ▢ $5 = 20 \div 4$

Part 13 **Write the fraction for each problem.**

a.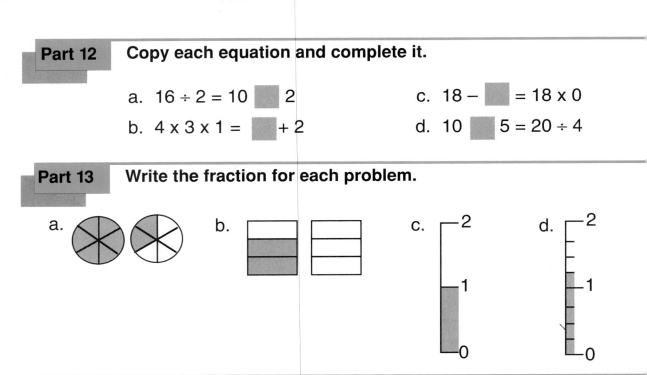

b.

c. -2 -1 -0

d. -2 -1 -0

Part 14 **Make a number family for each word problem. Then write the number problem and the answer.**

a. A dog ate 36 more ounces than a cat ate. If the cat ate 8 ounces, how many ounces did the dog eat?

b. There are 56 children in a room. 24 of the children are boys. How many of the children are girls?

Part 15 **Write the answer to each problem.**

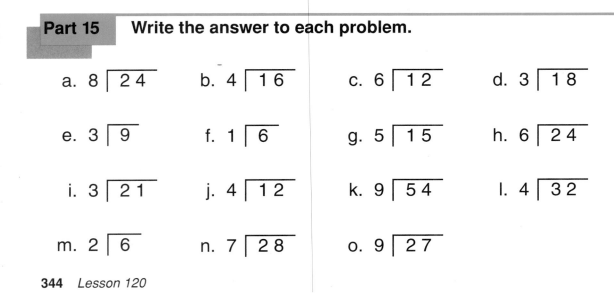

a. $8 \overline{\smash{)}24}$

b. $4 \overline{\smash{)}16}$

c. $6 \overline{\smash{)}12}$

d. $3 \overline{\smash{)}18}$

e. $3 \overline{\smash{)}9}$

f. $1 \overline{\smash{)}6}$

g. $5 \overline{\smash{)}15}$

h. $6 \overline{\smash{)}24}$

i. $3 \overline{\smash{)}21}$

j. $4 \overline{\smash{)}12}$

k. $9 \overline{\smash{)}54}$

l. $4 \overline{\smash{)}32}$

m. $2 \overline{\smash{)}6}$

n. $7 \overline{\smash{)}28}$

o. $9 \overline{\smash{)}27}$

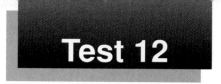

Part 5 — Copy the table. Then fill in the missing times and answer the questions.

	Time left	Minutes of trip	Time arrived
a. Fran	5:09		5:46
b. Ana		:41	7:56
c. Dan	5:19	:12	
d. Diane			
e. Roxanne			

d. Diane left for the party at 5:15. The trip took 38 minutes. When did she arrive at the party?

e. Roxanne left for the party at 5:12. She arrived at 5:31. How long did the trip take?

Part 6 — For each fraction write the division problem and the answer.

a. $\dfrac{24}{4}$ b. $\dfrac{50}{10}$ c. $\dfrac{35}{7}$

Part 7 — Make the number family. Write the column problem and figure out the answer.

a. If a woman spends 32 dollars more than her husband spends, and if the woman spends 234 dollars, how much does her husband spend?

b. If a ship weighs 12 tons more than a truck, and if the truck weighs 21 tons, how much does the ship weigh?

c. If the tank holds 350 more gallons than the barrel, and if the barrel holds 186 gallons, how many gallons does the tank hold?

Copy each problem and work it.

a. 9 ⟌ 5 0
    ⟌_____ R

b. 2 ⟌ 7
    ⟌_____ R

c. 4 ⟌ 1 1
    ⟌_____ R

Part 9 This table is supposed to show how many pounds 1 of each item weighs and 5 of each item weighs. Copy the table and fill in the missing numbers.

	Pounds for 1	Pounds for 5
Cans		20
Bottles		35
Jars		30

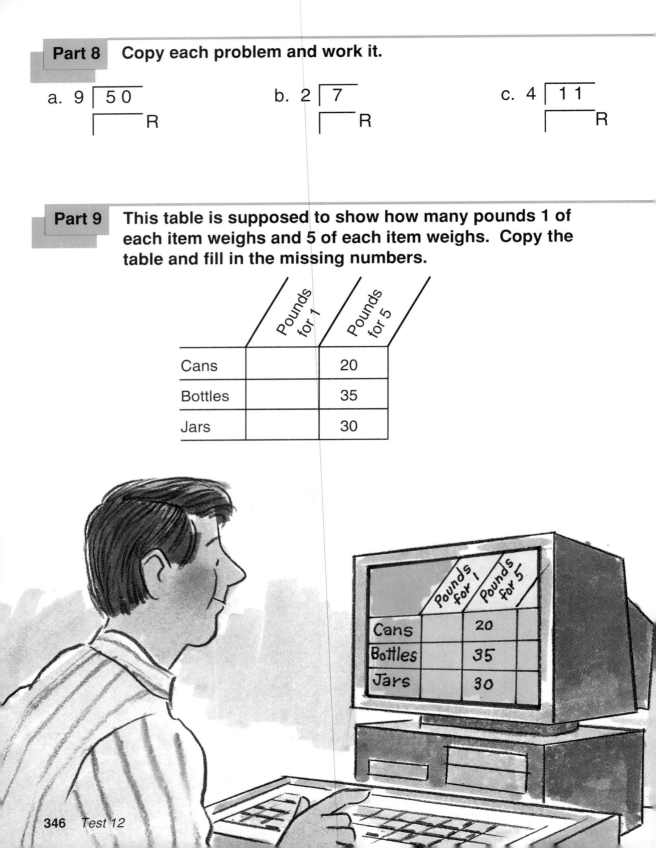

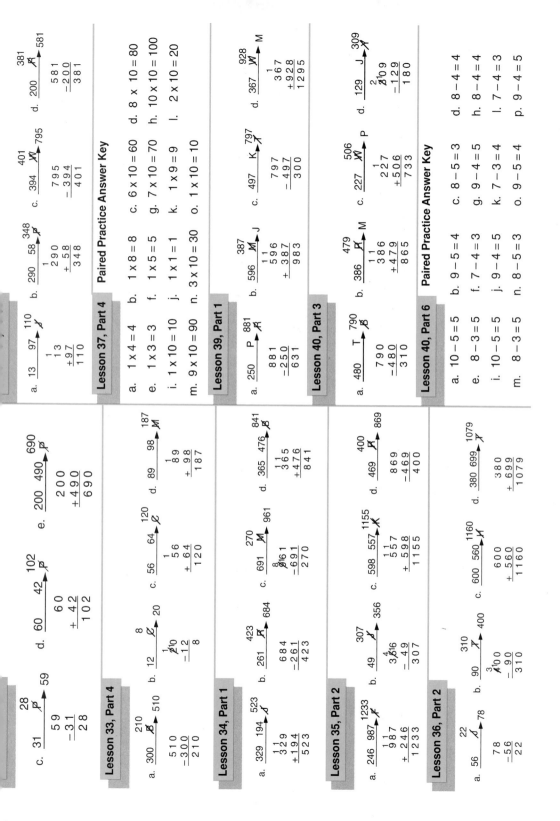

Lesson 33, Part 4

c. 31 → P 59
```
  59
 -31
  28
```
d. 60 → 42 → P 102
```
  60
 +42
 102
```
e. 200 → 490 → P 690
```
 200
+490
 690
```

a. 300 → B 510
```
 510
-300
 210
```
b. 12 → C 20
```
  1
 2̶1̶0
 -12
   8
```
c. 56 → 64 → C 120
```
  1
 56
+64
120
```
d. 89 → 98 → M 187
```
  1
 89
+98
187
```

Lesson 34, Part 1

a. 329 → 194 → A 523
```
  11
 329
+194
 523
```
b. 261 → R 423 → 684
```
 684
-261
 423
```
c. 691 → M 270 → 961
```
   8
 9̶6̶1
 -691
  270
```
d. 365 → 476 → B 841
```
  11
 365
+476
 841
```

Lesson 35, Part 2

a. 246 → 987 → F 1233
```
   1
  987
 +246
 1233
```
b. 49 → B 307
```
   4
 3̶5̶6
  -49
  307
```
c. 598 → 557 → X 1155
```
    1
  557
 +598
 1155
```
d. 469 → A 400 → 869
```
 869
-469
 400
```

Lesson 36, Part 2

a. 56 → A 78 → 22
```
  78
 -56
  22
```
b. 90 → A 310 → 400
```
   3
 4̶0̶0
  -90
  310
```
c. 600 → 560 → H 1160
```
  600
 +560
 1160
```
d. 380 → 699 → A 1079
```
  380
 +699
 1079
```

a. 13 → 97 → 110
```
   1
  13
 +97
 110
```
b. 290 → 58 → P 348
```
   1
 290
 +58
 348
```
c. 394 → 401 → M 795
```
 795
-394
 401
```
d. 200 → 381 → A 581
```
 581
-200
 381
```

Lesson 37, Part 4

a. 1×4=4   b. 1×8=8   c. 6×10=60   d. 8×10=80
e. 1×3=3   f. 1×5=5   g. 7×10=70   h. 10×10=100
i. 1×10=10 j. 1×1=1   k. 1×9=9   l. 2×10=20
m. 9×10=90 n. 3×10=30 o. 1×10=10

Lesson 39, Part 1

a. 250 → P 881 → R
```
 881
-250
 631
```
b. 596 → 387 → M J
```
  11
 596
+387
 983
```
c. 497 → K 797 → R
```
 797
-497
 300
```
d. 367 → 928 → M
```
   1
 367
+928
1295
```

Lesson 40, Part 3

a. 480 → T 790 → B
```
 790
-480
 310
```
b. 386 → 479 → R M
```
  11
 386
+479
 865
```
c. 227 → 506 → M P
```
   1
 227
+506
 733
```
d. 129 → J 309 → A
```
   2
 3̶0̶9
 -129
  180
```

Lesson 40, Part 6

a. 10-5=5   b. 9-5=4   c. 8-5=3   d. 8-4=4
e. 8-3=5   f. 7-4=3   g. 9-4=5   h. 8-4=4
i. 10-5=5   j. 9-4=5   k. 7-3=4   l. 7-4=3
m. 8-3=5   n. 8-5=3   o. 9-5=4   p. 9-4=5

a. Ann is older than Debbie. → D → A
b. Doug runs less than Jerry. → D → J
c. Jan has more paper than Carol. → C → J
d. Tony is shorter than Jack. → T → J
e. Fran eats less than Don. → F → D

Lesson 41, Part 4

a. 469 − 276 = 193
b. 496 − 297 = 199
c. 741 − 509 = 232
d. 734 − 299 = 435

Lesson 42, Textbook Part 2

a. Brian is younger than Debby. → B → D
b. Sid has less money than George. → S → G
c. Fran grew more trees than Jack did. → J → F
d. Tony weighs less than Jan. → T → J
e. Carol ran farther than Dorothy. → D → C

Lesson 42, Workbook Part 2

a. 568 − 457 = 111
b. 857 − 458 = 399
c. 808 − 310 = 498
d. 678 − 409 = 269
e. 843 − 199 = 644

---

a. 4 → 5 → 9   e. 4 → 5 → 9   i. 4 → 7 → 11
b. 4 → 6 → 10   f. 4 → 5 → 9   j. 4 → 6 → 10
c. 4 → 7 → 11   g. 4 → 8 → 12   k. 4 → 5 → 9
d. 4 → 8 → 12   h. 4 → 6 → 10   l. 4 → 7 → 11

Lesson 43, Part 1

a. 508 − 327 = 181
b. 425 − 196 = 229
c. 850 − 490 = 360
d. 746 − 349 = 397

Lesson 44, Part 1

a. 8060 − 2250 = 5810
b. 3241 − 1331 = 1910
c. 7810 − 1900 = 5910
d. 6789 − 2890 = 3899

Lesson 45, Part 3

a. 130 → 98 → T
b. 11 → ∅ → T
c. 23 → G → 153
d. 36 → K → 148

Lesson 46, Part 2

a. 40 → H → B → 130
b. 11 → ∅ → T → 98
c. 23 → G → 153
d. 36 → K → 148

Lesson 47, Part 2

a. 17 → G → 38
b. 22 → A → 131
c. 14 → F → G → 49
d. 31 → J → 40

Lesson 48, Part 2

a. 47 → K → 246
b. 103 → H → G → 307
c. 29 → D → 318
d. 400 → B → G → 199

Lesson 69, Part 2

	Monday	Tuesday	Total for both days
Red birds	21	**78**	99
Yellow birds	54	31	**85**
Total birds	75	**109**	184

Lesson 55, Part 3 — Paired Practice Answer Key

a. $10 - 6 = 4$ b. $8 - 6 = 2$ c. $12 - 6 = 6$ d. $9 - 6 = 3$
e. $11 - 6 = 5$ f. $7 - 6 = 1$ g. $11 - 6 = 5$ h. $10 - 6 = 4$
i. $9 - 6 = 3$ j. $12 - 6 = 6$ k. $8 - 6 = 2$ l. $9 - 6 = 3$
m. $10 - 6 = 4$ n. $11 - 6 = 5$ o. $12 - 6 = 6$

Lesson 61, Part 3 — Paired Practice Answer Key

a. $5 \times 1 = 5$ b. $5 \times 4 = 20$ c. $5 \times 2 = 10$ d. $5 \times 5 = 25$
e. $5 \times 3 = 15$ f. $2 \times 5 = 10$ g. $5 \times 1 = 5$ h. $3 \times 5 = 15$
i. $4 \times 5 = 20$ j. $5 \times 5 = 25$

Lesson 64, Part 3 — Paired Practice Answer Key

a. $12 - 6 = 6$ b. $10 - 4 = 6$ c. $10 - 6 = 4$ d. $8 - 6 = 2$
e. $11 - 6 = 5$ f. $9 - 6 = 3$ g. $11 - 5 = 6$ h. $9 - 3 = 6$
i. $8 - 2 = 6$ j. $12 - 6 = 6$ k. $7 - 6 = 1$ l. $9 - 4 = 5$
m. $9 - 6 = 3$ n. $10 - 6 = 4$ o. $9 - 3 = 6$

Lesson 68, Part 5 — Paired Practice Answer Key

a. $13 - 9 = 4$ b. $10 - 4 = 6$ c. $12 - 6 = 6$ d. $11 - 7 = 4$
e. $8 - 4 = 4$ f. $9 - 3 = 6$ g. $11 - 6 = 5$ h. $10 - 5 = 5$
i. $12 - 8 = 4$ j. $10 - 6 = 4$ k. $9 - 6 = 3$ l. $9 - 5 = 4$
m. $9 - 4 = 5$ n. $10 - 6 = 4$ o. $8 - 6 = 2$ p. $14 - 10 = 4$

Lesson 49, Part 3 — Paired Practice Answer Key

a. $2 \times 7 = 14$ b. $2 \times 9 = 18$ c. $2 \times 4 = 8$ d. $2 \times 5 = 10$
e. $2 \times 10 = 20$ f. $2 \times 6 = 12$ g. $4 \times 2 = 8$ h. $8 \times 2 = 16$
i. $9 \times 2 = 18$ j. $5 \times 2 = 10$ k. $3 \times 2 = 6$ l. $2 \times 9 = 18$
m. $6 \times 2 = 12$ n. $2 \times 4 = 8$ o. $7 \times 2 = 14$ p. $2 \times 6 = 12$

Lesson 51, Part 3 — Paired Practice Answer Key

a. $8 - 6 = 2$ b. $10 - 6 = 4$ c. $7 - 5 = 2$ d. $13 - 9 = 4$
e. $4 - 2 = 2$ f. $4 - 3 = 1$ g. $12 - 8 = 4$ h. $12 - 10 = 2$
i. $12 - 12 = 0$ j. $11 - 7 = 4$ k. $5 - 4 = 1$ l. $7 - 4 = 3$

Lesson 52, Part 2

a. 21 — J — 56 (subtract)
b. 26 — J — M — 95 (subtract)
c. 21 — M — 56 (add)
d. 28 — T — P — 91 (add)
e. 18 — J — 90 (add)

Lesson 53, Part 3

a. 18 — B — 90 (subtract)
b. 45 — B — T — 70 (add)
c. 17 — B — 59 (subtract)
d. 90 — B — J — 175 (add)
e. 12 — R — 96 (subtract)

Lesson 54, Part 4 — Paired Practice Answer Key

a. $4 \times 4 = 16$ b. $3 \times 4 = 12$ c. $2 \times 4 = 8$ d. $1 \times 4 = 4$ e. $4 \times 2 = 8$ f. $2 \times 4 = 8$ g. $4 \times 1 = 4$ h. $4 \times 4 = 16$
i. $4 \times 2 = 8$ j. $2 \times 5 = 10$ k. $4 \times 4 = 16$ l. $2 \times 8 = 16$ m. $4 \times 3 = 12$ n. $2 \times 3 = 6$ o. $1 \times 3 = 3$ p. $8 \times 2 = 16$

Paired Practice Answer Key

a. $9 \times 9 = 81$ b. $2 \times 9 = 18$ c. $3 \times 9 = 27$ d. $4 \times 9 = 36$

e. $1 \times 9 = 9$ f. $7 \times 9 = 63$ g. $8 \times 9 = 72$ h. $5 \times 9 = 45$

i. $6 \times 9 = 54$ j. $3 \times 9 = 27$ k. $5 \times 9 = 45$ l. $6 \times 9 = 54$

m. $8 \times 9 = 72$ n. $7 \times 9 = 63$

Lesson 72, Part 5   Paired Practice Answer Key

a. $16 - 8 = 8$ b. $16 - 10 = 6$ c. $6 - 3 = 3$ d. $6 - 5 = 1$

e. $6 - 2 = 4$ f. $12 - 6 = 6$ g. $12 - 10 = 2$ h. $4 - 0 = 4$

i. $4 - 4 = 0$ j. $10 - 10 = 0$ k. $10 - 5 = 5$ l. $18 - 10 = 8$

m. $18 - 9 = 9$ n. $14 - 7 = 7$ o. $7 - 7 = 0$

Lesson 76, Part 3   Paired Practice Answer Key

a. $3 \times 4 = 12$ b. $5 \times 4 = 20$ c. $7 \times 4 = 28$ d. $9 \times 4 = 36$

e. $8 \times 4 = 32$ f. $6 \times 4 = 24$ g. $4 \times 4 = 16$ h. $10 \times 4 = 40$

i. $4 \times 9 = 36$ j. $4 \times 6 = 24$ k. $4 \times 8 = 32$ l. $4 \times 3 = 12$

m. $4 \times 10 = 40$ n. $4 \times 7 = 28$ o. $4 \times 2 = 8$ p. $4 \times 5 = 20$

Lesson 78, Part 2   Paired Practice Answer Key

a. $8 - 3 = 5$ b. $12 - 6 = 6$ c. $6 - 3 = 3$ d. $14 - 7 = 7$

e. $8 - 5 = 3$ f. $7 - 3 = 4$ g. $10 - 7 = 3$ h. $11 - 3 = 8$

i. $11 - 10 = 1$ j. $8 - 4 = 4$ k. $11 - 8 = 3$ l. $12 - 3 = 9$

m. $11 - 9 = 2$ n. $10 - 5 = 5$ o. $7 - 4 = 3$ p. $10 - 3 = 7$

Lesson 84, Part 2   Paired Practice Answer Key

a. $15 - 10 = 5$ b. $11 - 6 = 5$ c. $11 - 5 = 6$ d. $12 - 5 = 7$

e. $14 - 9 = 5$ f. $13 - 5 = 8$ g. $10 - 5 = 5$ h. $15 - 5 = 5$

i. $14 - 5 = 9$ j. $12 - 5 = 7$ k. $14 - 9 = 5$ l. $13 - 5 = 8$

Paired Practice Answer Key

a. $\begin{array}{r} 15 \\ -9 \\ \hline 6 \end{array}$ b. $\begin{array}{r} 13 \\ -7 \\ \hline 6 \end{array}$ c. $\begin{array}{r} 14 \\ -9 \\ \hline 5 \end{array}$ d. $\begin{array}{r} 13 \\ -6 \\ \hline 7 \end{array}$ e. $\begin{array}{r} 13 \\ -9 \\ \hline 4 \end{array}$ f. $\begin{array}{r} 14 \\ -6 \\ \hline 8 \end{array}$

g. $\begin{array}{r} 12 \\ -6 \\ \hline 6 \end{array}$ h. $\begin{array}{r} 15 \\ -6 \\ \hline 9 \end{array}$ i. $\begin{array}{r} 16 \\ -6 \\ \hline 10 \end{array}$ j. $\begin{array}{r} 14 \\ -8 \\ \hline 6 \end{array}$ k. $\begin{array}{r} 13 \\ -6 \\ \hline 7 \end{array}$ l. $\begin{array}{r} 13 \\ -7 \\ \hline 6 \end{array}$

Lesson 95, Part 3   Paired Practice Answer Key

a. $5 \times 6 = 30$ b. $5 \times 8 = 40$ c. $5 \times 10 = 50$ d. $5 \times 9 = 45$

e. $5 \times 5 = 25$ f. $6 \times 5 = 30$ g. $7 \times 5 = 35$ h. $9 \times 5 = 45$

i. $8 \times 5 = 40$ j. $4 \times 5 = 20$ k. $2 \times 5 = 10$ l. $3 \times 5 = 15$

m. $5 \times 7 = 35$ n. $5 \times 9 = 45$ o. $5 \times 8 = 40$

Lesson 96, Part 4   Paired Practice Answer Key

a. $16 - 7 = 9$ b. $17 - 10 = 7$ c. $15 - 8 = 7$ d. $14 - 6 = 8$

e. $14 - 7 = 7$ f. $15 - 7 = 8$ g. $16 - 7 = 9$ h. $13 - 6 = 7$

i. $13 - 8 = 5$ j. $15 - 6 = 9$ k. $17 - 8 = 9$ l. $15 - 8 = 7$

m. $13 - 6 = 7$ n. $11 - 8 = 3$ o. $11 - 5 = 6$

Lesson 97, Part 5   Paired Practice Answer Key

a. $8 \times 5 = 40$ b. $3 \times 5 = 15$ c. $5 \times 5 = 25$ d. $5 \times 7 = 35$

e. $5 \times 3 = 15$ f. $2 \times 5 = 10$ g. $4 \times 5 = 20$ h. $6 \times 5 = 30$

i. $9 \times 5 = 45$ j. $5 \times 6 = 30$ k. $7 \times 5 = 35$ l. $5 \times 8 = 40$

m. $5 \times 10 = 50$ n. $5 \times 4 = 20$ o. $6 \times 5 = 30$

Lesson 98, Part 4   Paired Practice Answer Key

a. $8 \times 5 = 40$ b. $9 \times 5 = 45$ c. $5 \times 5 = 25$ d. $6 \times 5 = 30$

e. $5 \times 4 = 20$ f. $5 \times 6 = 30$ g. $5 \times 8 = 40$ h. $5 \times 3 = 15$

i. $5 \times 7 = 35$ j. $5 \times 9 = 45$ k. $5 \times 10 = 50$ l. $5 \times 3 = 15$

m. $4 \times 5 = 20$ n. $8 \times 5 = 40$ o. $6 \times 5 = 30$ p. $7 \times 5 = 35$